THE ANALYSIS
OF CONTROLLED
SUBSTANCES

Analytical Techniques in the Sciences (AnTS)

Series Editor: David J. Ando, Consultant, Dartford, Kent, UK

A series of open learning/distance learning books which covers all of the major analytical techniques and their application in the most important areas of physical, life and materials sciences.

Titles available in the Series

Analytical Instrumentation: Performance Characteristics and Quality
Graham Currell, University of the West of England, Bristol, UK

Fundamentals of Electroanalytical Chemistry
Paul M.S. Monk, Manchester Metropolitan University, Manchester, UK

Introduction to Environmental Analysis
Roger N. Reeve, University of Sunderland, UK

Polymer Analysis
Barbara H. Stuart, University of Technology, Sydney, Australia

Chemical Sensors and Biosensors
Brian R. Eggins, University of Ulster at Jordanstown, Northern Ireland, UK

Methods for Environmental Trace Analysis
John R. Dean, Northumbria University, Newcastle, UK

Liquid Chromatography–Mass Spectrometry: An Introduction
Robert E. Ardrey, University of Huddersfield, Huddersfield, UK

The Analysis of Controlled Substances
Michael D. Cole, Anglia Polytechnic University, Cambridge, UK

Forthcoming Titles

Infrared Spectroscopy: Experimentation and Applications
Barbara H. Stuart, University of Technology, Sydney, Australia

THE ANALYSIS
OF CONTROLLED
SUBSTANCES

Michael D. Cole
Anglia Polytechnic University, Cambridge, UK

WILEY

Email (for orders and customer service enquiries): cs-books@wiley.co.uk
Visit our Home Page on www.wileyeurope.com or www.wiley.com

Reprinted November 2005

This publication is designed to provide accurate and authoritative information in regard to the
subject matter covered. It is sold on the understanding that the Publisher is not engaged in
rendering professional services. If professional advice or other expert assistance is required, the
services of a competent professional should be sought.

Other Wiley Editorial Offices

John Wiley & Sons Inc., 111 River Street, Hoboken, NJ 07030, USA

Jossey-Bass, 989 Market Street, San Francisco, CA 94103-1741, USA

Wiley-VCH Verlag GmbH, Boschstr. 12, D-69469 Weinheim, Germany

John Wiley & Sons Australia Ltd, 33 Park Road, Milton, Queensland 4064, Australia

John Wiley & Sons (Asia) Pte Ltd, 2 Clementi Loop #02-01, Jin Xing Distripark, Singapore 129809

John Wiley & Sons Canada Ltd, 22 Worcester Road, Etobicoke, Ontario, Canada M9W 1L1

Wiley also publishes its books in a variety of electronic formats. Some content that appears
in print may not be available in electronic books.

Library of Congress Cataloging-in-Publication Data

Cole, M. D. (Michael D.)
   The analysis of controlled substances / Michael D. Cole.
     p. cm. – (Analytical techniques in the sciences)
   Includes bibliographical references and index.
   ISBN 0-471-49252-3 (alk. paper) – ISBN 0-471-49253-1 (pbk. : alk. paper)
    1. Drugs of abuse–Analysis. I. Title. II. Series.

RS190.D77 C647 2003
615′.78–dc21

                                 2002193367

British Library Cataloguing in Publication Data

A catalogue record for this book is available from the British Library

ISBN 10: 0-471-49252-3 (H/B)   ISBN 13: 978-0-471-49252-8 (H/B)
ISBN 10: 0-471-49253-1 (P/B)   ISBN 13: 978-0-471-49253-5 (P/B)

Typeset in 10/12pt Times by Laserwords Private Limited, Chennai, India
Printed and bound in Great Britain by Antony Rowe Ltd, Chippenham, UK
This book is printed on acid-free paper responsibly manufactured from sustainable forestry
in which at least two trees are planted for each one used for paper production.

For Mum, Dad and Ania, for all of their love, support and encouragement

Contents

Series Preface   xi

Preface   xiii

Acronyms, Abbreviations and Symbols   xv

About the Author   xvii

1   Introduction to Drug Trends, Control, Legislation and Analysis   1

    1.1   Introductory Remarks   1
    1.2   International Legislation   2
    1.3   Controlled Substances in the United Kingdom   3
          1.3.1   Background to the Misuse of Drugs Act, 1971   3
          1.3.2   The Provisions of the Misuse of Drugs Act, 1971   3
    1.4   Controlled Substances in the United States   5
    1.5   Controlled Substances in Australia   5
    1.6   The Drug Chemist and Drug Analysis   5
    1.7   Quality Assurance in the Drugs Laboratory   9
    1.8   Presentation of Evidence in Court   10
    References   11

2   Amphetamine and Related Compounds   13

    2.1   Introduction   13
    2.2   Qualitative Identification of Amphetamines   15
          2.2.1   Sampling and Physical Description of Amphetamines 15
          2.2.2   Presumptive Testing of Amphetamines   18
          2.2.3   Thin Layer Chromatography of Amphetamines   19
          2.2.4   Definitive Identification of Amphetamines   20

2.3 Quantification of Amphetamines   25
2.4 Comparison and Profiling of Amphetamine Samples   31
    2.4.1 The Leuckart Synthesis of Amphetamine   32
    2.4.2 The Reductive Amination of Benzyl Methyl Ketone   32
    2.4.3 The Nitrostyrene Synthesis   33
    2.4.4 Impurity Extraction and Sample Comparison   34
References   36

3 The Analysis of LSD   37

3.1 Introduction   37
3.2 Qualitative Identification of LSD   38
    3.2.1 Sampling and Physical Description of LSD Blotter Acid   39
    3.2.2 Extraction of LSD Prior to Analysis   41
    3.2.3 Presumptive Testing for LSD   42
    3.2.4 Thin Layer Chromatography of Samples Containing LSD   43
    3.2.5 Confirmatory Tests for the Presence of LSD   43
References   47

4 *Cannabis sativa* and Products   49

4.1 Introduction   49
4.2 Origins, Sources and Manufacture of Cannabis   51
4.3 Analytical Sequence, Bulk and Trace Sampling Procedures   53
4.4 Qualitative Identification of Cannabis   54
    4.4.1 Identification of Herbal Material   55
    4.4.2 Identification of Other Materials   57
    4.4.3 Comparison of Cannabis Samples   65
References   72

5 Diamorphine and Heroin   73

5.1 Introduction   73
5.2 Origins, Sources and Manufacture of Diamorphine   74
5.3 Appearance of Heroin and Associated Paraphernalia   77
5.4 Bulk and Trace Sampling Procedures   78
5.5 Identification, Quantification and Comparison of Heroin Samples   79
    5.5.1 Presumptive Tests for Heroin   80
    5.5.2 Thin Layer Chromatography of Heroin Samples   81
    5.5.3 Gas Chromatographic–Mass Spectroscopic
           Identification of Heroin   83
    5.5.4 Quantification of Heroin Samples   87
    5.5.5 Comparison of Heroin Samples   92
References   95

6 Cocaine   **97**

6.1   Introduction   97
6.2   Origins, Sources and Manufacture of Cocaine   99
      6.2.1   Extraction and Preparation of Coca Paste   99
      6.2.2   Synthesis of Pure Cocaine   100
6.3   Qualitative Identification of Cocaine   100
      6.3.1   Presumptive Tests for Cocaine   100
      6.3.2   Thin Layer Chromatography   102
      6.3.3   Definitive Identification of Cocaine   103
6.4   Quantification of Cocaine   107
      6.4.1   Quantification of Cocaine by GC–MS   107
      6.4.2   Quantification of Cocaine by UV Spectroscopy   108
6.5   Comparison of Cocaine Samples   109
References   110

7 Products from *Catha edulis* and *Lophophora williamsii*   **113**

7.1   Introduction   113
7.2   Products of *Catha edulis*   113
      7.2.1   Identification, Quantification and Comparison
              of Khat Samples   115
      7.2.2   Comparison of Khat Samples   119
7.3   Products of *Lophophora williamsii*   119
      7.3.1   Physical Description and Sampling of Materials   120
      7.3.2   Presumptive Tests for Mescaline   121
      7.3.3   TLC Analysis of Mescaline   121
      7.3.4   HPLC Analysis of Mescaline   122
      7.3.5   GC–MS Analysis of Mescaline   124
      7.3.6   Comparison of Peyote Samples   124
References   125

8 The Analysis of Psilocybin and Psilocin from Fungi   **127**

8.1   Introduction   127
8.2   Identification of Psilocybin- and Psilocin-Containing Mushrooms   129
      8.2.1   Identification of Fungal Species from Morphological
              Characteristics   129
      8.2.2   Identification of Psilocin and Psilocybin Using
              Chemical Analysis   130
      8.2.3   Quantification of Psilocin and Psilocybin by HPLC   135
8.3   The Identification of Psilocybin- and Psilocin-Containing
      Fungi Using DNA Profiling   136
References   137

9 The Analysis of Controlled Pharmaceutical Drugs – Barbiturates and Benzodiazepines   **139**

    9.1 Introduction   139
    9.2 Analysis of Barbiturates and Benzodiazepines   141
        9.2.1 Extraction of Barbiturates and Benzodiazepines
              from Dose Forms   142
        9.2.2 Presumptive Tests for Barbiturates and Benzodiazepines   142
        9.2.3 TLC of Barbiturates and Benzodiazepines   143
        9.2.4 Confirmatory Analysis of Barbiturates
              and Benzodiazepines   146
        9.2.5 Quantification of Barbiturates and Benzodiazepines   149
    References   151

10 Current Status, Summary and Conclusions   **153**

    10.1 Current Status of Drug Analysis   153
    10.2 Conclusions   155

Appendices   **157**

1 Presumptive (Colour) Tests   157
2 Less-Common Controlled Substances   159

Responses to Self-Assessment Questions   **161**

Bibliography   **175**

Glossary of Terms   **179**

SI Units and Physical Constants   **183**

Periodic Table   **187**

Index   **189**

Series Preface

There has been a rapid expansion in the provision of further education in recent years, which has brought with it the need to provide more flexible methods of teaching in order to satisfy the requirements of an increasingly more diverse type of student. In this respect, the *open learning* approach has proved to be a valuable and effective teaching method, in particular for those students who for a variety of reasons cannot pursue full-time traditional courses. As a result, John Wiley & Sons, Ltd first published the Analytical Chemistry by Open Learning (ACOL) series of textbooks in the late 1980s. This series, which covers all of the major analytical techniques, rapidly established itself as a valuable teaching resource, providing a convenient and flexible means of studying for those people who, on account of their individual circumstances, were not able to take advantage of more conventional methods of education in this particular subject area.

Following upon the success of the ACOL series, which by its very name is predominately concerned with Analytical *Chemistry*, the *Analytical Techniques in the Sciences* (AnTS) series of open learning texts has been introduced with the aim of providing a broader coverage of the many areas of science in which analytical techniques and methods are now increasingly applied. With this in mind, the AnTS series of texts seeks to provide a range of books which will cover not only the actual techniques themselves, but *also* those scientific disciplines which have a necessary requirement for analytical characterization methods.

Analytical instrumentation continues to increase in sophistication, and as a consequence, the range of materials that can now be almost routinely analysed has increased accordingly. Books in this series which are concerned with the *techniques* themselves will reflect such advances in analytical instrumentation, while at the same time providing full and detailed discussions of the fundamental concepts and theories of the particular analytical method being considered. Such books will cover a variety of techniques, including general instrumental analysis, spectroscopy, chromatography, electrophoresis, tandem techniques, electroanalytical methods, X-ray analysis and other significant topics. In addition, books in the series will include the *application* of analytical techniques in areas such as environmental science, the life sciences, clinical analysis, food science, forensic analysis, pharmaceutical science, conservation and archaeology, polymer science and general solid-state materials science.

Written by experts in their own particular fields, the books are presented in an easy-to-read, user-friendly style, with each chapter including both learning objectives and summaries of the subject matter being covered. The progress of the reader can be assessed by the use of frequent self-assessment questions (SAQs) and discussion questions (DQs), along with their corresponding reinforcing or remedial responses, which appear regularly throughout the texts. The books are thus eminently suitable both for self-study applications and for forming the basis of industrial company in-house training schemes. Each text also contains a large amount of supplementary material, including bibliographies, lists of acronyms and abbreviations, and tables of SI Units and important physical constants, plus where appropriate, glossaries and references to literature sources.

It is therefore hoped that this present series of textbooks will prove to be a useful and valuable source of teaching material, both for individual students and for teachers of science courses.

Dave Ando
Dartford, UK

Preface

The control of drugs is an emotive issue and has been, and will continue to be, the subject of much debate. Many drugs have medical uses and these, and others, are also used for 'recreational' purposes. A large number are also subject to control at both national and international levels. Many are addictive and their use can sometimes result in antisocial behaviour. Furthermore, their use is often associated with significant health risks, where these are known. It is not the intention of this present book to debate the 'rights and wrongs' of drug control and use. While drugs remain controlled, it will be necessary, within the legal context, for the forensic scientist to carry out a number of types of analyses, including the following:

1. Determine whether or not a controlled substance is present.
2. Determine how much of the substance is present.
3. Determine, on occasion, the relationship of drug samples to each other.

Drug analysis is one of the areas of forensic science where it is necessary to carry out an analytical investigation, in this case to prove whether a controlled substance is present or otherwise. In order to achieve this, a number of analyses are required, which must conform to the highest scientific standards. It is the aim of this text to illustrate the analyses that must be undertaken and why, to explain the processes and their underlying chemistry, and to give the reader an insight into why each of the analyses is performed. The book is not exhaustive in describing all of the methods that are available – there is a huge body of scientific literature available, including research methods that have not yet found casework applications. The choice of method will depend upon the resources and equipment available to the analyst, the legislative system in which the analyst is working and the questions being asked. The first chapter outlines the legal context of the analyses, while each of the subsequent chapters describe methods which can be applied to individual classes of drugs. The methods that are described in this book have, however, been used by the present author and in many examples the data are taken from casework materials, with the methods being known to work. By applying the principles described, the analyst should arrive at sound findings in terms of a particular analysis.

It would not have been possible to write this book without the support and encouragement of a great number of people, including colleagues and friends from around the world, and my family. For this, I thank them all.

Mike Cole
Anglia Polytechnic University, Cambridge, UK

Acronyms, Abbreviations and Symbols

General

AU	absorbance unit
CNS	central nervous system
DNA	deoxyribonucleic acid
ENFSI	European Network of Forensic Science Institutes
FTIR	Fourier-transform infrared (spectroscopy)
GC-ECD	gas chromatography, employing electron-capture detection
GC-FID	gas chromatography, employing flame-ionization detection
GC–MS	gas chromatography–mass spectrometry
HIV	human immunodeficiency virus
HPLC	high performance liquid chromatography
i.d.	internal diameter
IR	infrared
LLE	liquid–liquid extraction
PPE	personal protective equipment
QA	quality assurance
SIM	selected-ion monitoring
SPE	solid-phase extraction
TIC	total ion current
TLC	thin layer chromatography
UNDCP	United Nations Drug Control Programme
UV	ultraviolet
d_f	film thickness (of chromatography column)
M_r	relative molecular weight

m/z	mass-to-charge ratio
R	correlation coefficient
R^2	coefficient of determination
R_f	retardation factor (or relative front value)
t_R	retention time (chromatography)
λ_{max}	wavelength of maximum absorption in a UV spectrum

Chemical Species

BMK	benzyl methyl ketone
BSTFA	bis(trimethylsilyl)trifluoroacetamide
CBD	cannabidiol
CBN	cannabinol
TMS	trimethylsilyl
THC	tetrahydrocannabinol
LSD	lysergic acid diethylamide
MDA	3,4-methylenedioxyamphetamine
MDEA	3,4-methylenedioxyethylamphetamine
MDMA	3,4-methylenedioxymethylamphetamine
MSTFA	N-methyl-N-(trimethylsilyl)-2,2,2-trifluoroacetamide
N,O-BSA	N,O-bis(trimethylsilyl)acetamide
ODS	octadecasilyl
PCP	phencyclidine
HFBA	heptafluorobutyric anhydride

About the Author

Michael D. Cole, B.A. (Hons) Cantab., Ph.D.

Michael Cole graduated from the University of Cambridge in 1986 with a degree in Natural Sciences. From there his career progressed when he obtained a Ph.D. in Natural Product Chemistry from the University of London in 1990, having studied both at this university and The Royal Botanic Gardens, Kew, UK. In 1990, he joined the staff of the Forensic Science Unit at the University of Strathclyde as a short-course tutor, from where he progressed to Director of the Unit in 2000. In July 2001, Michael was appointed Professor of Forensic Science at Anglia Polytechnic University, Cambridge, where he now heads the Department of Forensic Science and Chemistry.

In addition to university duties, Michael was chairman of the European Network of Forensic Science Institutes Working Group on Drugs and Lead Assessor for the Council for the Registration of Forensic Practitioners Drugs Section, and has undertaken drug-related forensic casework in the UK and overseas. He has published a number of papers on drug analysis, particularly in the area of methods of drug identification and profiling.

Teaching and training in forensic science has always interested Michael and he has developed a number of short courses and integrated lecture courses, with a particular emphasis on drug analysis. These have been delivered in the UK, Europe, North and South America and in the Far East. Michael is particularly keen to continue to develop the educational provision in this discipline.

Chapter 1

Introduction to Drug Trends, Control, Legislation and Analysis

Learning Objectives

- To appreciate the problem of increasing drug use.
- To be aware of the international legislation relating to drugs.
- To be aware of the legislation in relation to the control of drugs in the United Kingdom, the United States and Australia.
- To appreciate the role of the drugs chemist in drugs analysis.
- To understand the need for quality assurance in the drugs laboratory.
- To gain an understanding of the ways to facilitate evidence presentation in court.

1.1 Introductory Remarks

The problems associated with psychotropic drugs and controlled substances have been, and continue to be, the subject of much debate. Regardless of one's views, however, there remains the fact that a number of drugs are *controlled substances*. There is now a considerable body of evidence that the number of people using controlled substances for non-medical purposes is increasing. Data from the United Kingdom (Figure 1.1) is mirrored by that collected from the international community.

Within the legal and forensic science context, in order to prove that an offence has been committed, it is necessary to prove that a drug is present, and, if required, to determine the amount of the drug and its relationship to other samples. It is essential for those working in this area to understand how such analyses are

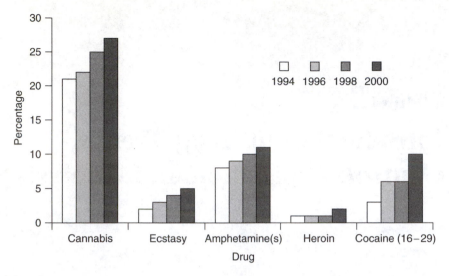

Figure 1.1 Percentage of 16–59 year-olds in the United Kingdom who claim to have used drugs – 'ever' (16–29 year-olds in the case of cocaine) [1].

carried out. In order to select, and critically evaluate, such analyses, it is also necessary to have an overview of the corresponding legislation in the jurisdiction in which one is working.

1.2 International Legislation

Within the international context, controls on drugs are set out in three treaties issued by the United Nations, namely:

1. The Single Convention on Narcotic Drugs, 1961.
2. The Convention on Psychotropic Substances, 1971.
3. The Convention against Illicit Traffic in Narcotic Drugs and Psychotropic Substances, 1988.

Signatories to these treaties implement control through domestic laws. In the United Kingdom, the principle legislative document for drug control is the *Misuse of Drugs Act, 1971*. This has been the subject of 14 modification orders and is accompanied by the *Misuse of Drugs Act (Regulations), 1985*, which was superceded by the *Misuse of Drugs Act (Regulations), 2001*.

Within the United States, the situation is further complicated because drugs are scheduled at the Federal level, but there may also be legislation at the State and County levels.

1.3 Controlled Substances in the United Kingdom

1.3.1 Background to the Misuse of Drugs Act, 1971

In the UK today, the legislative documents that are used to control drugs of abuse are the Misuse of Drugs Act, 1971, its amendments and modification orders, and the Misuse of Drugs Act (Regulations), 2001, which supersedes the Misuse of Drugs Act (Regulations), 1985. In essence, the Misuse of Drugs Act, 1971 defines what may not be done with respect to these compounds, while the Misuse of Drugs Act (Regulations), 2001 defines what may be done under the appropriately controlled circumstances. Similarly, customs offences, such as 'knowingly evading prohibition on unauthorized import/export of controlled substances', are regulated by the *Customs and Excise Management Act, 1979*.

Historically, in the United Kingdom the *Dangerous Drugs Act, 1951* simply controlled vegetable narcotics, such as *Cannabis sativa* (cannabis) and opium, and a few chemically related synthetic substances. This was superseded by the *Dangerous Drugs Act, 1964*, which organized the controlled drugs into three schedules based on internationally accepted principles. This was the first time that stimulants, used as anorectics, such as amphetamine and its analogues, were included in British Law. It also introduced some specific offences in relation to cannabis. In 1965, a new act, i.e. the *Dangerous Drugs Act, 1965*, combined the provisions of the Dangerous Drugs Act, 1951 with those of the Dangerous Drugs Act, 1964, as well as providing a more comprehensive definition of herbal cannabis as 'the fruiting and flowering tops of any plant of the genus *Cannabis*'. Since the forensic scientist still came across difficulties in discriminating fragmented plant parts which could still be a potent source of the active constituents of the plant, herbal cannabis was therefore redefined in the Misuse of Drugs Act, 1971 as: 'all the aerial parts, except the lignified stem and the non-viable seed, of any plant of the genus *Cannabis*'.

Another problem to be corrected by the Misuse of Drugs Act 1971 was that of the analogues of amphetamines, which were defined as: 'structurally derived by substitution in the side-chain or by ring closure therein' in the Act of 1965. Several compounds, such as ephedrine, were specifically excepted, but over 90 others were not purposely included. This was corrected by naming the specific compounds. Care was taken to re-phrase the wording so that certain chemical compounds, having the potential to become drugs of abuse, which might not yet have been available, generally referred to as 'designer drugs', would still be included. References were made, for example, to 'ether and ester derivatives' and to the 'stereoisomers' of several compounds.

1.3.2 The Provisions of the Misuse of Drugs Act, 1971

The Misuse of Drugs Act, 1971 lists controlled substances in three classes in Schedule 2 to the Act. Class A drugs have the greatest propensity to cause social harm, and Class C drugs the least. Class A drugs include cocaine, heroin, mescaline, morphine and opium, Class B includes amphetamine(s), and Class C the benzodiazepines. At the time of writing,[†] Cannabis is being reclassified. In addition, stereoisomers, salts, esters, ethers and certain preparations are also controlled groupwise, thus removing the need to name each of these individually. Associated with each class of drug are maximum penalties which may be prescribed. Those for Class A drug offences are more severe than those for Class C offences. For Class A drugs, some offences carry a maximum sentence of life imprisonment, for Class B 14 years in prison, and for Class C, five years in prison. With respect to each of the listed drugs, the Misuse of Drugs Act, 1971 is divided into several sections (Table 1.1), with each section relating to a specific type of offence under the Act which is prohibited.

In addition, the Government may create exceptions to the general rules and allow certain substances to be imported and exported, allow persons to use certain drugs under licence, allow medical and veterinary practitioners to supply certain drugs, and allow certain persons to manufacture, possess and work with drugs for educational or scientific research purposes. The mechanism by which much of this is achieved is detailed in the Misuse of Drugs (Regulations), 2001, which details what *may be done* and *how*, while the Misuse of Drugs Act, 1971 details what *may not be done*.

In this legal area, it is necessary to be able to provide scientific support for any charge brought against individuals to prove that an offence has been committed. The majority of offences relate to possession of controlled substances. However, it is sometimes necessary for the analyst to determine the amount as well as the

Table 1.1 Principle sections of the Misuse of Drugs Act, 1971 and corresponding offences

Section of the Misuse of Drugs Act, 1971	Type of offence which is controlled
3	Importation and exportation of controlled drugs
4	Production and supply of controlled drugs
5	Possession of controlled drugs
6	Cultivation of cannabis
8	Permit premises to be used for the purposes listed in Sections 3, 4, 5, 6 and 9
9	Preparing or smoking opium
9	Use utensils or allow others to do so in relation to smoking opium
20	Induce the commission of a 'corresponding offence' while overseas

[†] May, 2002.

presence of a drug and on occasion, particularly in relation to supply offences, establish relationships between drug samples. The amount of work that is required depends upon the drug in question and the charge being made. For a small amount of heroin, for personal use, and on admission of guilt, sufficient support is offered by a colour (presumptive) test. However, if the admission is later retracted, a full scientific investigation of the drug is required. For other drug types, it is possible to prove the identity by the simple use of microscopy. This is especially true for cannabis products and the identification of some fungi. However, for other case types a full and rigorous investigation must be undertaken.

1.4 Controlled Substances in the United States

At the Federal level, controlled substances are listed within a system of five schedules in the *Controlled Substances Act*. These Schedules are described in Table 1.2. Schedule I contains the most strongly controlled substances, while Schedule V includes the most moderately controlled. Those drugs contained in Schedules II to V may be prescribed, while those in Schedule I may not. The data in the table illustrate a point which requires to be addressed, particularly at cross-border (International, State or County) levels, that is, one of nomenclature. In the United Kingdom, 'heroin' is taken to mean the mixture of products resulting from the synthesis of diamorphine from morphine. Both compounds are listed separately in UK legislation, although 'heroin' is not. However, in the United States, 'heroin' can sometimes be taken to mean diamorphine and the two are sometimes used interchangeably.

1.5 Controlled Substances in Australia

In a situation analogous to that in the United States of America, legislation covering drugs of abuse has been written at the Territory and Commonwealth levels. The two principle documents relevant at Commonwealth level are the *Customs Act, 1901* and the *Crimes (Traffic in Narcotic Drugs and Psychotropic Substances) Act, 1990*.

1.6 The Drug Chemist and Drug Analysis

Forensic science, the application of scientific principles to the legal process, is especially important in drugs analysis because in every case, one or more samples must be investigated in order to prove, or otherwise, that a controlled substance is present. The drugs chemist must ensure that the materials provided are suitable for the analysis to be carried out, select the correct materials, carry out the correct analysis and achieve quality data of a certain standard, interpret

Table 1.2 Federal scheduling of controlled substances in the United States of America

Schedule	Examples	Potential for abuse	Acceptance or otherwise for medical use	Safety of drug under medical supervision
I	Lysergic acid diethylamide (LSD), 3,4-methylene-dioxymethyl-amphetamine (MDMA), cannabis, psilocybin, heroin	High	No currently accepted medical use	Lack of accepted safety data
II	Cocaine, morphine, opium, amphetamine, phencyclidine (PCP)	High	Accepted medical use with severe restrictions	Abuse may lead to severe psychological or physical dependence
III	Ketamine, lysergic acid, marinol (synthetic tetrahy-drocannabinol (THC))	Potential less than Schedules I and II drugs	Medical use accepted	Abuse may lead to moderate/low psychological or physical dependence
IV	Benzodiazepines	Low potential for abuse, cf. Schedules I, II or III	Medical use accepted	Abuse may lead to limited psychological or physical dependence relative to Schedule III
V	Prescription medicines containing low doses of codeine, etc.	Low potential for abuse, cf. Schedules I, II, III and IV	Medical use accepted	Abuse may lead to limited psychological or physical dependence relative to Schedule IV

the findings and present them in written and/or verbal form. Forensic scientists, and drug analysts in particular, should think of themselves as witnesses for the court – not specifically for the prosecution or defence. Their objective is to assist the court to *reach* decisions about either the innocence or guilt of the accused.

The majority of this text deals with specific drug classes, but regardless of the legislative system one is working in, or the drug class in question, a number of basic forensic science principles should be followed at all times. The basic analytical process follows the sequence shown in Figure 1.2.

Having received the sample into the laboratory, the drug analyst should consider the particular question(s) being asked and whether the relevant answers can be obtained from the sample which has been provided. If the answer is in the affirmative, he/she should then proceed. The item should be fully documented and described, including the condition of the packaging. If for any reason this is not intact, the analysis should not go ahead. The data on the label should also be recorded and the analyst should sign and date the label, to ensure that continuity of evidence is complete. All of these data should be recorded contemporaneously, in a system in which each page is contiguously numbered. The analyst should also sign every page, and each sheet of paper that is produced by any instruments used throughout the course of the analysis.

Having recorded all of the physical data available, the decision must then be made as to whether the item contains trace or bulk materials. The latter can be

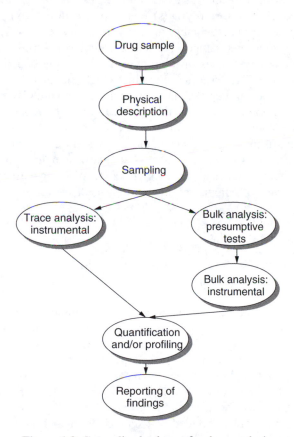

Figure 1.2 Generalized scheme for drug analysis.

seen with the naked eye and present the opportunity to contaminate other samples. Trace samples cannot be seen and/or are easily contaminated. Examples include the surfaces used to cut drugs, knives and the surfaces of balances used to weigh the materials. Given that contamination is an important issue, the analyst should not handle trace samples if bulk samples have previously been handled or if the analyst has been exposed to bulk samples. Ideally, trace samples should be analysed in a separate laboratory which is demonstrably free from contamination. Protective clothing is especially important if (i) trace samples are being handled, and (ii) *very* large samples are being examined. The former situation is important because it prevents the sample itself becoming contaminated, and the latter because it reduces the risk of contact with or ingestion of the drug by the analyst. Personal protective equipment (PPE) or clothing should be used/worn whenever required, particularly since some of the reagents used to analyse drugs are caustic or corrosive and many of them are classified as 'harmful'.

If the specimen provided is a trace sample, sufficient material should be recovered to allow an instrumental analysis directly. The nature of the sample will often provide a clue as to the drug(s) involved and direct comparison can be made by using gas chromatography–mass spectrometry (GC–MS), for example. If the specimen is a bulk sample, presumptive (colour) tests are undertaken to determine the class or classes of drugs which the sample contains. Thin layer chromatography (TLC) is used to determine which members of the classes are present and it might also be possible to make a semi-quantitative estimate of the amount(s) of drug(s) present. Standard mixes can then be prepared for use in the confirmatory techniques.

If drug comparisons are to be made, the drugs themselves and also the wrapping/packaging materials may be compared. It is now recognized by the European Network of Forensic Science Institutes (ENFSI) Working Group (Drugs) that there are four levels of comparison of the drug itself, as follows:

1. Drug identification
2. Drug quantification
3. Identification of cutting agents added to the drug
4. Chemical impurity profiling

The first two of these processes are the same as for simple drug identification and quantification. Identification of the cutting agents may be of help, for example, where a rare cutting agent is used, but this information cannot be used to definitively state that two drug samples are related. This latter can only be achieved by examination and analysis of the chemical impurities arising from the drug manufacture and preparation itself. Most drugs are produced in a batchwise process and each batch will have a unique profile which will not be affected by the addition of cutting agents, provided that the cutting agent(s) does not appear in the impurity profile.

Having carried out the analysis, whether it be a qualitative, quantitative or comparative analysis, a report must be written. There is no universal 'recipe' for such report writing. However, there is some essential information which must be included, namely the full name, age (may be stated as over 21 on Witness Statements in England and Wales) and qualifications of the person(s) preparing the report. The time (date) on which the items were received and from whom must be recorded. The materials analysed should be detailed. Much of this latter information is taken directly from the physical description which will have been previously prepared. The report should then state the findings and conclusions of the analyst. This includes the major facts, for example, what the drug is, how much is present, and in which legislative body it is controlled. The drug analyst, as a forensic scientist, is also able to express *evidence of opinion* (unlike other witnesses in court) and hence can express a view on whether two or more samples were related, how many doses a certain amount of drug might form, etc. What should always be remembered, however, is that the expert witness should never stray outside his own particular area of expertise.

The report may, or may not, contain a materials and methods section – opinion is currently divided on this point. Certainly, the technical section (materials and methods) should be at the end of the report, as this will mean very little to a lawyer unless the latter is well trained in science. It is therefore better to present such information at the end where interested parties can read the details should they desire. In terms of content, it is common, where such sections are included in reports, to simply list common or widely accepted methods that are used while at the same time detailing new methods, or variations on an accepted method, in some depth, so that another scientist may repeat the work.

Finally, before presentation of the report, the format should be checked – double spacing (to allow annotation), wide left-hand margins (to allow binding into thick documents), and single-sided printing (to facilitate reading) should all be ensured. Every page should be contiguously numbered and each page should be signed and dated by the person preparing the report – even if it is only a proforma with check boxes, as used in some jurisdictions.

1.7 Quality Assurance in the Drugs Laboratory

In order for the report and evidence presented in court to be of value, a regular programme of quality assurance must be entered into by the analyst and his laboratory. Records of the methods used should be available, along with documented assessments of the performances of each of the tests and instruments used. Such records are essential should the functionality of the tests or equipment be challenged.

Any new method to be used in the laboratory should be rigorously tested, to ensure that reproducibility, repeatability and robustness comply with internationally accepted standards. It should also have been through a peer-reviewing process and ideally be widely accepted by the relevant scientific community.

In terms of ensuring that performance standards are met, it is preferable for the laboratory to participate in proficiency tests – both declared and undeclared. There are a number of these commercially available, or they may be prepared in-house, although this latter approach is always open to accusations of 'results fixing' and bias. While many analysts are understandably nervous about such tests, when properly handled they can be used to improve the performance of the laboratory.

1.8 Presentation of Evidence in Court

The job of the drug analyst would not be complete without the *presentation of a report*. On occasion, this may be in verbal (in addition to written) form in the witness box. The processes which occur will vary with where the evidence is being given, but certain guidelines will make the situation less traumatic. The analyst should be well prepared and know the case. The documentation should be complete and preferably bound in a file, and again, the analyst should know this file intimately. Many of the common questions can be anticipated once experience is gained, but these can be guessed at in an educated way at an early stage. It is advisable to have as much experience as possible prior to giving evidence in court and to have observed lawyers and scientists in action, either for real, or in a moot court.

At court, be smartly dressed and be punctual. Adopt a good posture in the witness box. In giving the answers to the questions, be precise and accurate, without being technical. If the answer is not known, this should be stated. If there are attempts by the legal practitioners to mislead, confuse or misstate your evidence, remember that the judge is there to correct these misconceptions. With diligent application in the laboratory and in the courtroom, the materials will have been correctly analysed and the findings successfully reported.

Summary

There is now a considerable body of evidence that the problem of drug use is increasing. Attempts to control drug use and abuse are made at the international level through United Nations legislative documents, which are mirrored in signatory states by legislation at the national and sub-national levels.

In the UK, the principle legislative documents are the Misuse of Drugs Act, 1971, plus its amendment orders, and the Misuse of Drugs (Regulations), 2001. In the United States, drugs are controlled at the Federal level by the Controlled Substances Act. Control also exists at the State and County levels. An analogous position is found in Australia, where drugs are controlled at both Commonwealth and Territory levels.

The role of the forensic scientist within this legal framework is to assist the court in deciding whether or not a drug offence has been committed. This is achieved by establishing whether or not a controlled substance is present, how much is present and, on occasion, the relationship (or otherwise) of the material being considered to other drug samples. A scheme of work should be followed, depending upon whether the material is a trace or a bulk sample and quality assurance measures should always be rigorously followed.

Findings should always be reported in a clear and concise manner, which can be understood by the layman. This is particularly important when oral evidence is presented in court, although the same principles also apply to written evidence. Technical evidence should sometimes be included, but not at the expense of clarity.

References

1. Drugscope, *Annual Report on the UK Drug Situation, 2001* [www.drugscope.org/druginfo/drugreport.asp].[†]

[†] Accessed, November 2002.

Chapter 2

Amphetamine and Related Compounds

Learning Objectives

- To be aware of the common types of amphetamines on the illicit drug market.
- To learn the methods available for the qualitative identification of amphetamines.
- To appreciate the methods available for the quantification of amphetamines.
- To know the methods available for the profiling of amphetamines.

2.1 Introduction

There are a large number of amphetamines which are controlled substances. Of these, the most commonly encountered in the forensic science laboratory are amphetamine (**1**), methylamphetamine (**2**), 3,4-methylenedioxyamphetamine (MDA) (**3**), 3,4-methylenedioxymethylamphetamine (MDMA) (**4**) and 3,4-methylenedioxyethylamphetamine (MDEA) (**5**) (see Table 2.1). In addition, there are a wide variety of structurally related analogues which can be synthesized [1]. The usage pattern of the drugs is regional, and subject to temporal variation. A number of the drugs, particularly those related to amphetamines, are stimulants, while the 3,4-methylenedioxy derivatives also induce a feeling of well-being and in addition act as entactogens. The drugs are most commonly ingested orally, but can also be insufflacated, smoked or injected. The dose required depends upon the drug being used, the degree of use by the user and the required effect. Some typical doses are presented in Table 2.1. When ingested orally, the drugs are frequently taken in tabletted form. When smoked,

Table 2.1 Structures and typical doses of commonly encountered amphetamines

Drug	Typical dose
1	Up to 30–50 mg
2	Up to 30–50 mg
3	100–150 mg
4	75–125 mg
5	120–150 mg

the drug material is purchased in powdered form and then inhaled by using special apparatus.

The drugs themselves can be totally synthetic, and indeed, precursor chemicals such as benzyl methyl ketone are subject to separate controls and monitoring. Alternatively, they may be semi-synthetic, where natural products have been used as the starting materials for the synthesis. There are a number of synthetic routes described in the literature [1, 2], in addition to a plethora of Internet sites where recipes are described.

The drugs themselves are mixed with a wide variety of adulterants and diluants and also contain a cocktail of related products and impurities from the manufacturing process. These products can be used for drug comparison but the effects of these on the drug user are not fully understood and may be extremely harmful.

DQ 2.1

Why are adulterants and diluants added to a sample?

Answer

*There are a number of reasons for these additions. **Adulterants**, for example, caffeine and other pharmacologically active drugs, are added to either hide the lack of the desired drug, dilute it (and hence increase profit for the drug dealer), or add another type of effect to the drug mixture. **Diluents** are bulking agents and may include starch, talc, etc., and are added to make the drug 'go further'. Some additives may be added to the powders, either to improve the flow properties of the tablets prior to the tabletting process, or to impart a particular colour.*

In the United Kingdom, at the time of writing,[†] the amphetamines are Class B drugs under Schedule 2 of the Misuse of Drugs Act, 1971, unless they are prepared for injection, in which case they become Class A drugs. The following details the analysis of the most common of this drug class, although the principles can be applied to all of the related compounds that might be encountered in a forensic science laboratory.

2.2 Qualitative Identification of Amphetamines

A wide variety of types of sample may be expected by the forensic scientist. These include powdered drug material, tabletted material and items likely to have traces of the drug samples present. From these items, the appropriate samples should be taken, described and subsequently forwarded for analysis.

2.2.1 Sampling and Physical Description of Amphetamines

2.2.1.1 Powder Samples

If the samples are powdered materials, they should be sorted into groups, where the members of the groups cannot be distinguished from each other. Having achieved this, the items in each group should be counted and a good physical description prepared. This should include weight, colour, odour and any other physical characteristic that the scientist considers to be important. Depending upon the number of items in the group, the following sampling strategy, based upon the United Nations Drug Control Programme (UNDCP) recommendations, may be adopted. If there are between 1 and 10 items, all of them should be examined. If there are between 10 and 100 items, then 10 items should be examined, while if there are more than 100 items, the square route of the number of items

---

[†] May, 2002.

should be examined. These should be chosen at random from among the items, by, for example, assigning each of them a number and then choosing the items to be examined from a random-number table.

DQ 2.2

What advantage is there in using a random-number table or generator to determine which of the samples should be analysed?

Answer

The advantage of this is that it removes any bias which might be imparted by the operator in sample choice.

Other sampling protocols exist, but a discussion of these is beyond the scope of this present text. There are, however, a number of problems associated with sampling procedures in general. If, for example, the sample comprises 1000 items, then a large number of expensive and time-consuming analyses will be required. If it is assumed that there is only one drug type in the seizure, then recent work using 'probabilistic reasoning' has shown that it may be possible to analyse as four or six individual items, depending upon the degree of certainty required [3]. However, the difficulty arises when more than one type of drug or drug mixture is present, but the different mixtures are visually identical. Whether this occurs or not depends upon the area in which one is working. In some areas, the problem (in amphetamines, at least) is reported not to arise [4], while in parts of mainland Europe, it is known to exist [5]. As a consequence, such a model, where reduced numbers are sampled, may only be applicable after a great deal of work to establish that multiple visually identical drug types do not occur within seizures. In essence, the number of items to be analysed depends upon the drug in question, the experience of the analyst and the jurisdiction in which he is working.

Once the samples to be analysed have been identified, physical descriptions, homogenization and sampling, presumptive testing, TLC and confirmatory testing should be undertaken.

2.2.1.2 Tabletted Samples

If the items are tabletted, then a different approach is required. The items should be divided into visually indistinguishable groups and the number of items in each group should be counted, if necessary estimating the number from the mean weight of tablet and the total mass of the tablets, if very large numbers are involved. A good physical description of the tablets should be made, including recording of the size, colour, shape, logo and score marks (if present), while the ballistics (physical characteristics) of the tablet should be examined and detailed. Photography is particularly helpful in this respect. This latter includes recording of all of the physical damage to the tablets which may be present.

> **SAQ 2.1**
>
> What is the advantage of recording the ballistic features of tabletted drug units?

Having recorded all of the physical data, the items to be examined chemically should be chosen. At the time of writing,[†] there is no agreed best-practice protocol for undertaking this and the analyst should work within the requirements of the judicial system in which he/she is operating. The theory applied to powdered amphetamines can be equally applied to amphetamine tablets. The latter should be sampled and presumptive tests, TLC and confirmatory tests then carried out.

2.2.1.3 Trace Samples of Amphetamines

If the samples are trace samples, that is, the drug present is likely to be easily contaminated, the approach should be that the operators, laboratory equipment and reagents to be used should be demonstrably free of drug residues prior to the commencement of the analysis. This can be achieved by washing the glassware, work surfaces and operator's hands with a small amount of methanol and concentrating the dissolved materials. The same batch of solvent should be used for this procedure and for the analysis of the drug items themselves. The control extract should be analysed in the same run sequence as the materials from the case samples (it is not usually possible to carry out the control analyses prior to commencing further work because of time and cost constraints). If the controls are drug-free, then any drug observed in the casework samples can only have come from the latter materials themselves.

Having carried out the control procedures, the item(s) to be examined should then be swabbed, individually, by using a clean swab soaked in a suitable solvent. A suitable solvent should freely dissolve the drug material, not cause decomposition of the drug or react with it, and be amenable to subsequent analytical procedures. It should be remembered that the swabbing process should leave enough material intact for a second and subsequent analysis. If the swab is not to be used immediately, it should be dried and stored until needed.

> **SAQ 2.2**
>
> Why should the swab be dried prior to storage if it is not going to be used immediately?

2.2.1.4 Sample Homogenization

Homogenization of powdered samples can be achieved by using a number of different methodologies. One of the best methods for samples likely to be encountered 'on the street' is the use of the 'cone-and-square' method. In this technique,

---

[†] May, 2002.

the materials are mixed and the larger fragments reduced in size. The material is poured onto a flat, clean surface and then divided into four; opposite quarters are removed and the two remaining quarters recombined. The process is repeated until the desired sample size is achieved. For larger samples, a 'core' may be taken and then subjected to further homogenization by using this cone-and-square methodology.

For tabletted materials, homogenization is more problematic since if the tablet was homogenized, it would also be destroyed. For this reason, a sample is taken from the tablet and gently scraped from the dose form, away from any ballistic features which are likely to be of use in subsequent examinations. The powder is then thoroughly homogenized prior to testing.

SAQ 2.3

Why is it important to retain the ballistics features where possible?

2.2.2 Presumptive Testing of Amphetamines

Having described the materials and sampled them, it is then necessary to determine which of the amphetamines might be present. Of the presumptive tests available, the most commonly used is the Marquis test (see Appendix 1). In this approach, a small amount of the 'Marquis reagent' is added to the case sample(s), while in addition, positive and negative controls are also performed. The Marquis test yields yellow to orange colour reactions with amphetamine and methamphetamine, depending upon the concentration of the drug in the sample – the more drug is present, then the more intense and orange the colour reaction. For the 3,4-methylenedioxy compounds, a purple–blue colour is obtained, which should not be confused with the almost indigo colour that is obtained when opiates are tested with the Marquis reagent.

DQ 2.3

How can this confusion which may arise about which drug is present be overcome?

Answer

This is easily achieved by undertaking positive controls when the presumptive test is attempted on the sample material. Such controls provide direct reference points against which colour changes can be matched.

Having established that an amphetamine might be present in the sample, the next step is to determine which of this group of controlled substances is present. This can be rapidly achieved by using TLC.

2.2.3 Thin Layer Chromatography of Amphetamines

In order that the sample can be tested for the presence of amphetamines, a test solution must be prepared. The sample should be dissolved in a suitable solvent (methanol is commonly used) at a sample concentration of the order of 10 mg ml^{-1}. This allows for the fact that many amphetamine samples at the 'street level' are extremely weak, i.e. between 2 and 10% amphetamine in a matrix of adulterants and diluants, giving a solution of approximately $0.2–1.0 \text{ mg ml}^{-1}$, namely a concentration at which the standards can be prepared.

SAQ 2.4

Why is methanol the preferred solvent for the preparation of samples for TLC in amphetamine analysis?

SAQ 2.5

What is the problem with using methanol as a solvent in the TLC analysis of amphetamines?

The sample should be dissolved as fully as possible and centrifuged or filtered to remove any solid particulates. A positive and negative control should also be prepared. The silica gel chromatographic plate should be marked up and the test solutions, plus the positive and negative controls, placed on the plate and the latter allowed to develop in the chosen solvent system (Table 2.2).

DQ 2.4

What is the function of the ammonia in the methanol/acetone/ammonia mixture (see Table 2.2)?

Table 2.2 Typical TLC data (R_f/values) obtained for some common amphetamines

Compound	Solvent system	
	MeOH/Me$_2$CO/NH$_3$ (25/6/0.4, by vol)[a]	CHCl$_3$/MeOH (4/1, by vol)[b]
Amphetamine	0.62	0.22
Methylamphetamine	0.26	0.23
MDA	0.60	—
MDMA	0.24	—
MDEA	0.37	—

[a]Data taken from reference [6].
[b]Data taken from reference [7].

Answer

*The ammonia is added to achieve a process known as **ion suppression**. By converting the drugs to their free base forms, their polarities are reduced. This is because the nitrogen atom does not carry a positive charge in basic solution. The latter reduces the problem of (TLC) tailing, improves the mass transfer properties between the stationary and mobile phases, and thus improves the chromatographic quality.*

Having developed the chromatogram, the plate should be removed from the solvent tank, dried at room temperature and then examined under (i) white light, (ii) short-wavelength UV light (254 nm), (iii) long-wavelength UV light (360 nm), and (iv) after spraying with a developing reagent. One method which has been found to be particularly successful is lightly spraying the chromatogram with 0.5 M NaOH, allowing the plate to dry and then spraying with a solution of 0.5% (wt/vol) Fast Black K in water. Amphetamine produces a purple colour, while methylamphetamine gives an orange/red colour. In addition, MDA, MDMA and MDEA give rise to purple, orange/red and orange/red products, respectively [6]. At each of the visualization stages, the retardation factor (or relative front) (R_f) values of the visualized compounds should be calculated by using the following equation:

$$R_f = \frac{\text{Distance moved by the analyte of interest}}{\text{Distance moved by the solvent front}} \tag{2.1}$$

The R_f values of the unknowns are compared to those of the standards and if the data cannot be discriminated then a suggested match is called.

Although when using this combination of presumptive tests and TLC it is possible to discriminate within this group of compounds, due to the extremely large number of amphetamines available, it is necessary to carry out a confirmatory analytical technique. The foremost of these, for amphetamine identification, is gas chromatography–mass spectrometry (GC–MS).

2.2.4 Definitive Identification of Amphetamines

GC–MS is the preferred method for the identification of amphetamines. The discussion below centres on the analysis of amphetamine itself, although the same principles can also be applied to other members of this class of drug. However, there are a number of problems associated with the gas chromatographic analysis of amphetamine. Being highly polar in nature, this compound is liable to poor chromatographic behaviour and tailing if the analytical instrument is not scrupulously clean (Figure 2.1).

In addition, while in some GC–MS systems it is possible to obtain a molecular ion and individual fragments for amphetamine (Figure 2.2) under the majority of conditions, amphetamines without ring substitutions fragment in a very similar

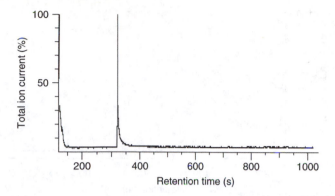

Figure 2.1 Gas chromatogram of amphetamine obtained from a GC–MS analysis, showing the tailing that can occur during the analysis of this compound.

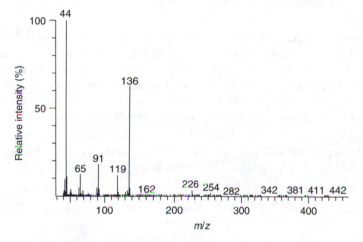

Figure 2.2 Electron-impact mass spectrum of amphetamine.

way, sometimes producing little or no molecular ions, plus common fragments at *m/z* 119, 91, 65 and 51.

Furthermore, the highly polar nature of the amino group results in sorption of amphetamine to the surfaces of the GC system components. This, coupled with the often low concentration of the amphetamine in the sample, results in the false impression that there is no amphetamine present in the specimen under investigation (Figure 2.3).

In order to alleviate this problem, derivatization can be employed. One of the easiest processes, for the analysis of amphetamine, is to derivatize directly with carbon disulfide and it is this method which finds wide application in the United Kingdom. For bulk and trace samples, this is achieved by dissolving the material

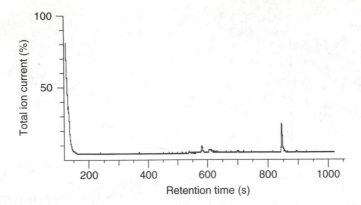

Figure 2.3 Gas chromatogram of a case sample of amphetamine where the latter has disappeared from the trace due to sorption onto the GC system components. Note that the caffeine in the sample is still recorded (t_r, 845 s).

in the swab (trace samples) or approximately 2 mg of the specimen (bulk samples) in 0.1 ml ammoniacal methanol and adding 0.2 ml carbon disulfide to the mixture, which should be thoroughly shaken and then allowed to stand, at room temperature, for at least 30 min. The mixture should be evaporated to dryness under a stream of nitrogen and the mixture resuspended in methanol, to which an appropriate internal standard has been added. It should be remembered that a positive and negative control should also be prepared. The mixtures can then be chromatographed in the following order – negative control, blank, positive control, blank, sample blank..., etc. – with the necessary check standards as appropriate.

SAQ 2.6

Why is the amphetamine dissolved in ammoniacal methanol (e.g. 100 ml of methanol containing 200 µl of 880 ammonia)?

DQ 2.5

What is the reaction of amphetamine with carbon disulfide (CS_2) and what does it achieve?

Answer

The reaction (see equation (2.1)) is a simple, pre-column derivatization, involving the amino group of the amphetamine and the CS_2. This process reduces the polarity of the product, improving its chromatographic behaviour and hence the sensitivity of the method. In addition, it results *in a molecule which produces characteristic fragments from the ioniza-tion process*:

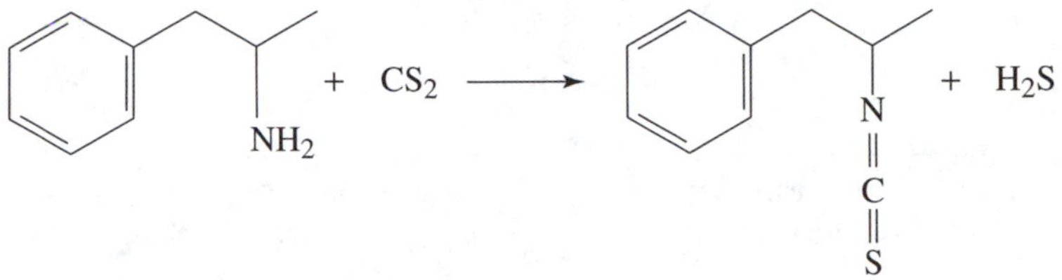

$$+ \quad CS_2 \longrightarrow \qquad\qquad + \quad H_2S \tag{2.2}$$

The conditions employed for a suitable GC analysis of such mixtures are presented in Table 2.3.

Derivatization results in good chromatographic behaviour and provides compounds for which distinctive mass spectra can be obtained (Figure 2.4).

This method also has the advantage that the most common adulterant of amphetamine, namely caffeine, can also be identified in this system, and a suitable mass spectrum obtained (Figure 2.5). As with all drug identification approaches, if the retention time data and mass spectra of the compounds match, then an identification can be inferred.

SAQ 2.7

Is it possible to explain the origins of the fragments observed in the mass spectrum of amphetamine derivatized with carbon disulfide (see Figure 2.4(c))?

One difficulty with this method is that the derivatization process will not work for secondary amines such as methylamphetamine and 3,4-methylenedioxy methylamphetamine (MDMA). An alternative method for derivatization is therefore required in such cases. One such technique is the derivatization of the amphetamine samples with heptafluorobutyric anhydride (HFBA) [8].

Table 2.3 GC operating conditions and parameters suitable for the analysis of amphetamine derivatized with carbon disulfide

System/parameter	Description/conditions
Column	BP-5: 25 m $\times$ 0.22 i.d.; d_f, 0.5 μm
Injection temperature	280°C
Column oven temperature programme	100°C for 1 min; increased to 240°C at 12°C min^{-1}; held for 5 min
Carrier gas	He, at a flow rate of 1 ml min^{-1}
Split ratio	10:1

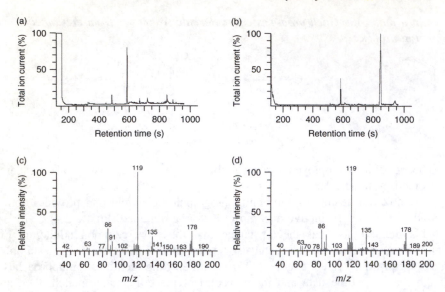

Figure 2.4 GC–MS data obtained for the NCS derivatives of standard and street samples of amphetamine: (a) gas chromatogram of standard amphetamine derivatized with CS_2 (t_r, 585 s); (b) gas chromatogram of a street sample of amphetamine derivatized with CS_2 (t_r (amphetamine), 585 s; t_r (caffeine), 845 s); (c) mass spectrum of the NCS derivative of standard amphetamine; (d) mass spectrum of the NCS derivative of amphetamine in a street sample.

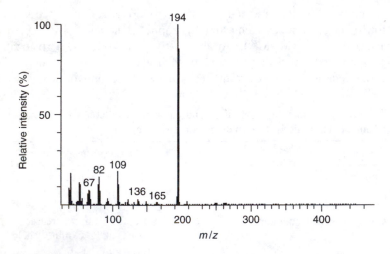

Figure 2.5 Electron-impact mass spectrum of caffeine following GC–MS separation of a street sample of amphetamine.

In this system, amphetamine, 3,4-methylenedioxyamphetamine (MDA) and the N-methyl and N-ethyl analogues of the drugs were derivatized, with 4-chloroamphetamine being used as the internal standard. For powders, solutions were prepared at concentrations of 1 mg ml^{-1} in methanol and a 100 μl aliquot blown dry under nitrogen. To the residue was added 50 μl of 0.5 M KOH and 500 μl of toluene. The mixture was then shaken for 30 s and centrifuged, after which the toluene layer was recovered and HFBA (5 μl) added. The excess reagent was neutralized with 500 μl of 10% NaHCO$_3$ and an aliquot (1 μl) analysed by GC–MS. Each of the amphetamines and the N-alkylated derivatives could be separated and identified on the basis of retention index and mass spectral data.

SAQ 2.8

What are the structures of the derivatives when amphetamine and methylamphetamine react with heptafluorobutyric anhydride (HFBA)? The base ion of the HFBA derivative of amphetamine occurs at *m/z* 254. How does this arise?

By matching retention time data and mass spectra, the compounds can be identified. Having identified the actual compounds present, it may then be necessary to quantify them.

2.3 Quantification of Amphetamines

Due to the nature of the compounds being considered and the need for derivatization, GC–MS is not considered the best technique for sample quantification.

DQ 2.6

What are the difficulties associated with quantification following derivatization?

Answer

There are a number of difficulties encountered with quantification after employing derivatization. These include the fact that we are now considering another analytical step, with the concomitent increase in cost and time of each analysis. In addition, because derivatization is another handling stage in the analytical process, there is always the risk of sample contamination. Furthermore, the assumption is made that the derivatization reactions are 'complete' and that the corresponding derivatives are stable for the period between derivative formation and analysis. Further factors are that dilutions need to be extremely accurate and precise to obtain reliable numerical data and that derivatization can potentially lead to increases in numerical errors for such data.

While capillary electrophoretic methods have been employed and do find some application in casework [9, 10], the method of choice is currently high performance liquid chromatography (HPLC). The exact method to be employed will depend upon the amphetamine in question, but the method described below [11] can be applied to amphetamine, methylamphetamine, MDA and MDMA without any difficulties.

The amphetamines (standards and samples) should be dissolved in methanolic HCl (100 ml of methanol to which 175 μl of concentrated HCl has been added). A range of standard solutions should be prepared in order to give a range of concentrations above and below that which the street sample is thought to contain, remembering that the latter may only contain between 0 and 5 wt% amphetamine. If necessary, the materials (particularly the case samples) should be sonicated and, following this, centrifuged to remove any solid materials. The supernatant is retained for subsequent analysis.

DQ 2.7

Why is the sample dissolved in methanolic HCl?

Answer

This is carried out in order to form the hydrochloride salt form of the drug. This is because that in this system the separation process is based upon, among other mechanisms, an ion-exchange process.

The operating conditions for a typical HPLC system which can be used in such analyses are given in Table 2.4. While this system has been found to work efficiently, there are other HPLC methods available for the analysis of amphetamine, particularly with perchlorate buffers [7]. However, the general principles remain the same. When setting this method into operation, the HPLC column should be allowed sufficient time to equilibrate with the mobile phase – typically 30–60 min.

SAQ 2.9

Why is this long equilibration time required?

The calibration standards should be analysed from the weakest to the strongest, with blanks between samples to demonstrate that there has been no 'carry-over' between injections. The samples can then be analysed, with 'check standards' analysed between every 4th or 5th sample.

SAQ 2.10

Why should the calibration curve be prepared from the weakest sample to the strongest?

Table 2.4 HPLC operating conditions and parameters suitable for the quantitative analysis of amphetamines

System/parameter	Description/conditions
Column	Silica gel: 12.5 cm × 4.6 mm i.d.; spherical 5 μm particles
Mobile phase	MeOH/NH$_3$/HCl (2000:18.4:5.8)[a]
Flow rate	1 ml min^{-1}
Injection volume	5 or 10 μl (injection loop)
Detection	UV at 254 nm, with a diode-array detector

[a] By volume.

Table 2.5 Chromatographic and spectroscopic data for the commonest amphetamines and caffeine obtained by using the HPLC system described in the text

Compound	λ$_{max}$ (nm)	Retention time (min)
Caffeine	273	1.19
Amphetamine	259	1.68
MDA	235, 287	1.69
Methylamphetamine	260	2.67
MDMA	235, 287	2.76

SAQ 2.11

What are 'check standards' and what is their function?

The retention time and λ$_{max}$ data for the commonest amphetamines, plus caffeine, are given in Table 2.5. While it is not possible to fully resolve the pairs, amphetamine and MDA, and methylamphetamine and MDMA, on the basis of their retention times, discrimination between these compounds can be achieved by using a diode-array detector (Figure 2.6).

Having collected the data, a calibration curve should be plotted. Since amphetamine is frequently synthesized in dirty apparatus in 'clandestine' laboratories, it may not be possible to determine which salt form of the drug is present. The standard is generally supplied as the sulfate form, of the general formula (amphetamine)$_2$ sulfate. This means that for every gram of amphetamine sulfate, 73% will be present as the amphetamine free base. The calibration curve should be plotted as (UV detector) response against concentration of amphetamine free base. Exemplar data are presented in Table 2.6 and Figure 2.7.

The results should be checked to ensure that the calibration data do truly lie on a straight line. Any spurious data points should be discarded and the remainder used to formulate the calibration equation. The method of *least-squares* is one technique which can be used, where the following simultaneous equations:

$$\sum Y = nc + m \sum X \tag{2.3}$$

$$\sum XY = c \sum X + m \sum X^2 \tag{2.4}$$

are solved, for the data generated. Alternatively, a computer spreadsheet can be used. These data yield a *regression equation* of $Y = 0.0304X + 0.000\,04$, where X is the concentration of amphetamine as free base and Y is the detector response (see Figure 2.7 and Table 2.6).

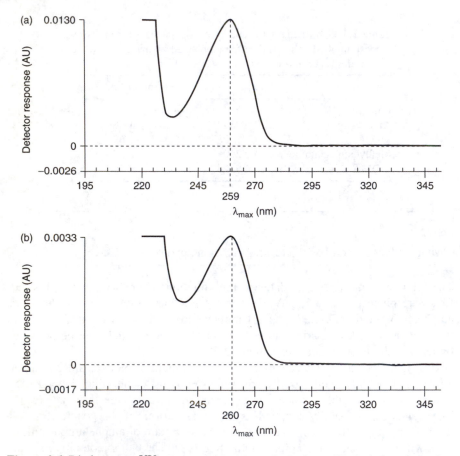

Figure 2.6 Diode-array UV spectra of (a) amphetamine, (b) methylamphetamine, (c) MDA and (d) MDMA, following HPLC analysis using the system described in the text.

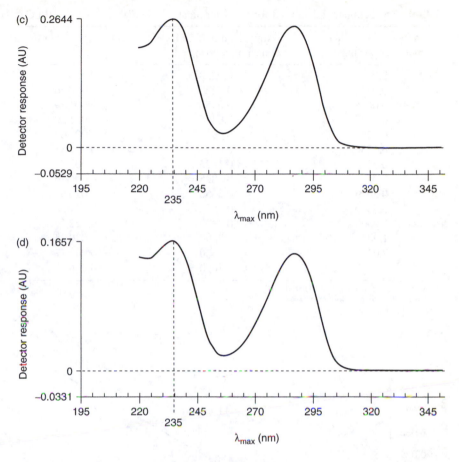

Figure 2.6 (*Continued*).

The equation generated can then be used as follows. If a street sample, prepared at 2.5 mg of sample per ml, gives a reading of 0.003 14 AU,[†] then the concentration of free base is 0.102 mg ml^{-1} (obtained by substitution of the Y-value – the absorbance reading – into the regression equation). However, a correction is required to account for the original concentration of the sample and so the actual concentration, expressed as a percentage, then becomes $(0.102/2.5) \times 100 = 4.08\%$ amphetamine as the free base.

DQ 2.8

Is it necessary to calculate a regression equation for every sample?

[†] AU, absorbance unit.

Table 2.6 Calibration data used for the determination of amphetamine by HPLC

Amphetamine sulfate concentration (mg ml^{-1})	Amphetamine free base concentration (mg ml^{-1})[a]	Detector response[b] (AU)[c]
0.0313	0.0228	0.000 74
0.0313	0.0228	0.000 66
0.0625	0.0456	0.001 79
0.0625	0.0456	0.001 50
0.1250	0.0913	0.003 07
0.1250	0.0913	0.002 81
0.2500	0.1825	0.005 78
0.2500	0.1825	0.005 76
0.5000	0.3650	0.010 59
0.5000	0.3650	0.011 05
1.0000	0.7300	0.022 76
1.0000	0.7300	0.020 94
2.0000	1.4600	0.042 77
2.0000	1.4600	0.046 65

[a] X-value (see text for further details).
[b] Y-value (see text for further details).
[c] AU, absorbance units.

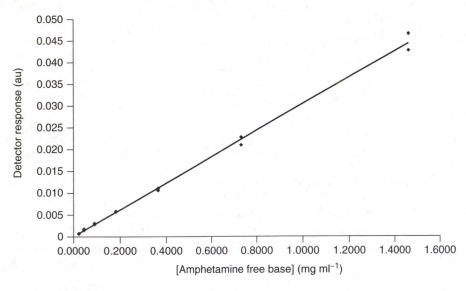

Figure 2.7 A typical calibration curve used for the determination of amphetamine by HPLC.

Answer

*The short answer to this is no. Linearity and a linear dynamic range (the range of concentrations over which the detector response is linear) must be achieved when the method is first developed. From this point on, a range of standards within this linear range can be used to calculate a regression equation. The latter can then be employed to obtain the concentration of the sample being analysed. Some laboratories use standards of known concentration to determine if there is deviation from the line (this is equivalent to a single point estimate being used as a check standard), while others use low- and high-concentration standards to recalculate the equation of the line every five or ten samples. This latter method need not impose a great deal of work on the operator as computers can be programmed to carry out this function. Both approaches are valid. The important point is to ensure that this function is carried out. However, if any components of the analytical system are changed, it is **then** necessary to recalibrate the instrument before proceeding with further analysis.*

2.4 Comparison and Profiling of Amphetamine Samples

A great deal is known about the numerous synthetic routes available for the manufacture of amphetamines. These substances are produced in batchwise processes, with each batch containing a different impurity profile comprising different amounts of decomposition products, reaction by-products and unchanged starting materials. As a consequence, the different impurity profiles can be used as signatures which can be used to establish a link between samples or otherwise. Even when the same route is used twice, different impurity profiles are present in the samples. This allows several types of investigation to proceed. Small street samples may be related to each other, as may larger batches in transit. In addition, chemists in different countries prefer different routes of synthesis. Thus, with the right intelligence, patterns for different syntheses in different countries can be established and the country of origin identified. The three principle routes of synthesis are the Leuckart synthesis, the reductive amination of benzyl methyl ketone (BMK) and the nitrostyrene route. It is not the intention here to provide the 'recipes' for these syntheses – they may be readily obtained from the refereed scientific literature and the Internet, although brief outlines are given below. What is intended is to illustrate the different types of impurities which can provide signatures when different routes are used, to discuss the extraction and analysis of the impurities, and the interpretation of the data following such analyses.

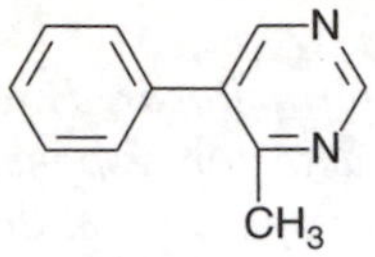

Figure 2.8 Exemplar impurities found in amphetamine produced by using the Leuckart synthesis.

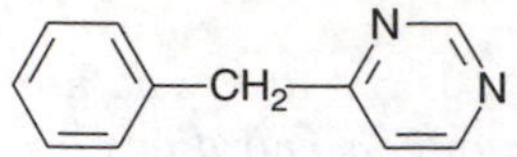

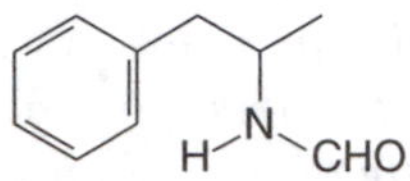

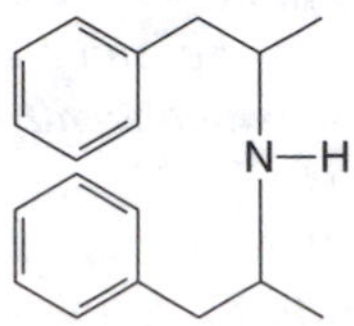

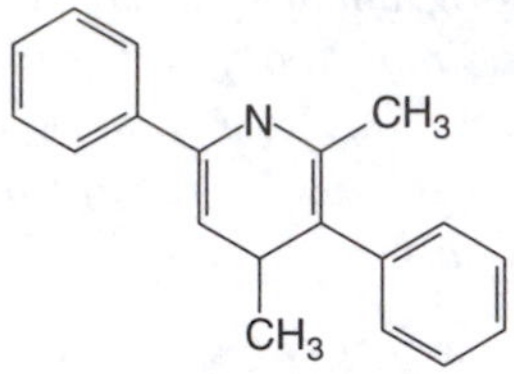

2.4.1 The Leuckart Synthesis of Amphetamine

In this reaction, benzene methyl ketone (BMK) is reacted with formamide, followed by reaction of the products with hydrogen peroxide and HCl in methanol, to form amphetamine. There are a number of variations on the basic route. The impurities which arise include, as examples, methylphenylpyrimidines (4-methyl-5-phenylpyrimidine) (**8**), benzylpyrimidines (4-benzylpyrimidine) (**9**), *N*-formylamphetamine (**10**), *N*,*N*-di(β-phenylisopropyl)amines (*N*,*N*-di(β-phenylisopropyl)amine) (**11**) and dimethyldiphenylpyrimidines (2,4-dimethyl-3,5-diphenylpyridine) (**12**) (see structures in Figure 2.8).

These impurities lead to complex profiles, but ones which are extremely characteristic of the synthetic route being used and, frequently, of the chemist who has been manufacturing the drug [12].

2.4.2 The Reductive Amination of Benzyl Methyl Ketone

There are a number of variations on this route, depending upon how the reduction is catalysed. The general principle is that the ketone is reacted with ammonia to form the imine, which is then reduced under high temperature and pressure to form the amine. Examples of impurities from this route include ketimines

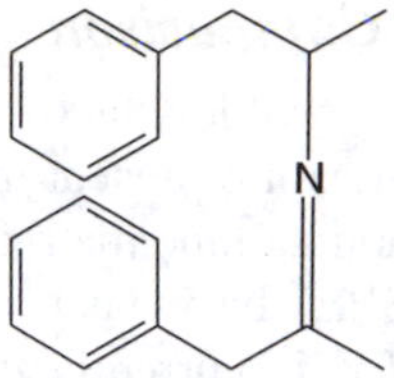

13

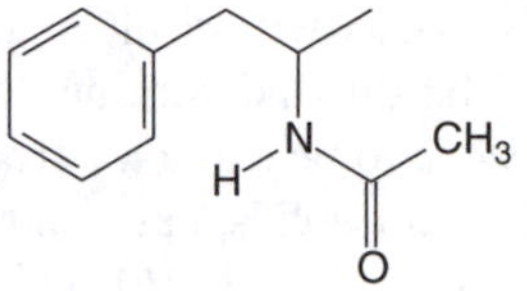

14

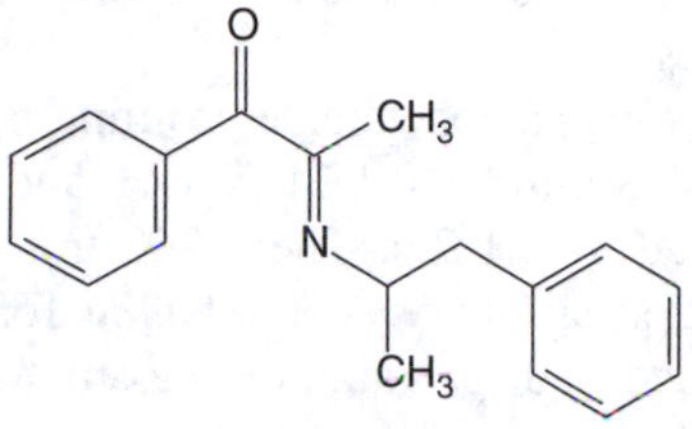

15

Figure 2.9 Exemplar impurities found in amphetamine synthesized by reduction amination of BMK.

(N-(β-phenylisopropyl)benzyl methyl ketimine) (**13**) and N-acetylamphetamine (**14**), plus more complex reaction by-products such as 1-oxo-1-phenyl-2-(β-phenylisopropylimino)propane (**15**) (see structures in Figure 2.9).

2.4.3 The Nitrostyrene Synthesis

This third route is not as widely encountered as those described above. However, a number of characteristic impurities are found in amphetamine samples synthesized via this process, including nitrostyrene (**16**), benzyl methyl ketoxime (**17**) and 1-methyl-2-phenylaziridine (**18**) (see structures in Figure 2.10).

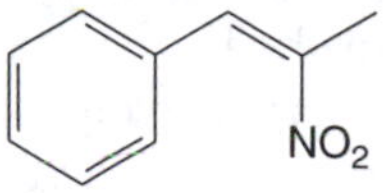

16

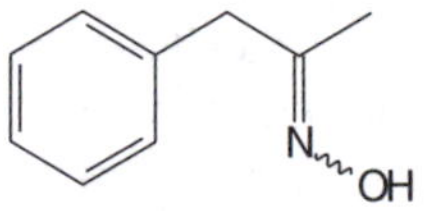

17

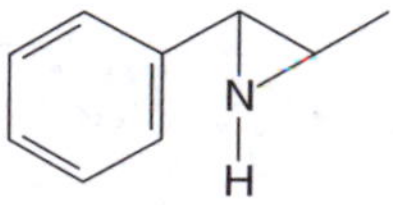

18

Figure 2.10 Exemplar impurities found in amphetamine synthesized via the nitrostyrene route.

2.4.4 Impurity Extraction and Sample Comparison

Each route of synthesis will result in an impurity profile which can be analysed. Unlike cannabis, heroin and cocaine, the impurities are present in very low concentrations and need to be extracted from the amphetamine matrix. Liquid–liquid extraction (LLE) and solid-phase extraction (SPE) have been most frequently applied for this purpose [13–16]. Of these, LLE is currently the most widely used technique. In addition, there are attempts being made to develop the method so that it can be optimized to allow data exchange between different laboratories (in different countries) [17].

The method upon which amphetamine profiling is currently based is as follows. A buffer solution of 30 ml of 0.1 M NaOH and 50 ml of 0.1 M Na_2PO_4 is prepared and adjusted to pH 7.0. Then, 200 mg of the drug are dissolved in 2 ml of buffer solution and the system shaken for 30 min, after which 200 μl of *n*-hexane are added and the mixture shaken for a further 30 min. The two phases are allowed to separate – this can be enhanced by using centrifugation. A small aliquot (say 2 μl) of the sample can then be analysed by using either gas chromatography, employing flame-ionization detection (GC-FID) or GC–MS.

DQ 2.9

How does the LLE system work?

Answer

Such a system exploits the differences in the pK_a values of amphetamine and the various impurities. The latter have pK_a values close to 7 and so are not charged at this pH level. The amphetamine remains charged and as a consequence, remains water-soluble. The impurities can therefore be extracted into the organic phase and concentrated, being selectively removed from the amphetamine in which they were found.

Since the analysis of amphetamine is so complex and a definitive statement must be made about the route of synthesis if profiling is employed, the technique of choice for analysis is GC–MS. A variety of methods have been published but in general these all require a slow temperature programme on a moderately polar capillary GC column. An example of a typical impurity profile obtained by using GC–MS is shown in Figure 2.11.

DQ 2.10

Why is it necessary to use GC–MS for amphetamine profiling?

Answer

This is needed because the mixtures, as shown in Figure 2.11, are extremely complex, often containing more than 50 compounds. In

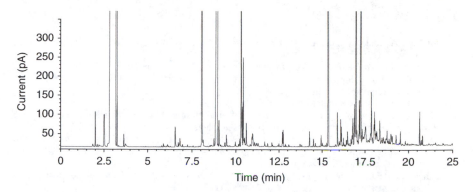

Figure 2.11 Impurity profile obtained from an amphetamine sample by using GC–MS.

addition, the identities of many of these are not known. The only instrumentation offering the required resolution, and the ability to compare the identities of the various compounds in different samples, is GC–MS. This is the reason why this hybrid technique is the method of choice in such analyses.

The comparison can then begin. The simplest way to do this is to 'lay the chromatograms over each other' and decide whether they are the same or not. Alternatively, there are a number of algorithms which can be used, either retrospectively or predictively, to establish relationships between samples. However, such an approach is currently at the research stage and no definitive method has yet been identified. The method used will ultimately depend upon the question(s) being asked and the desired outcome of the use of the data.

Summary

There are now a large number of amphetamine-related products which are subject to drug control. Items for analysis may include powders, tablets, liquids and the various paraphernalia associated with drug taking. Of particular importance are the ballistic features of tablets which may allow comparison of samples to be made. The items should be described and where necessary, suitable samples taken for analysis. For bulk drugs, this entails presumptive testing, TLC and GC–MS for quantification, with extraction of synthetic intermediates, by-products and decomposition products for GC–MS analysis when drug profiling. For trace samples, analysis is directly provided by GC–MS.

Drug profiling has been carried out on a wide range of amphetamines manufactured by a variety of different routes. Information about these synthetic routes, laboratory networks and transport and distribution networks can all be obtained provided that the data are handled in the appropriate way.

References

1. Shulgin, A. and Shulgin, A., *Phenethylamines I Have Known and Loved*, Transform Press, Berkeley, CA, 1991.
2. Buzz, P., *Recreational Drugs*, Loompanics Unlimited, Port Townsend, WA, 1989.
3. Aitken, C. G. G., 'Sampling – how big a sample?', *J. Forensic Sci.*, **44**, 750–760 (1999).
4. Jamieson, A., personal communication.
5. Lock, E., personal communication.
6. Munro, C. H. and White, P. C., 'Evaluation of diazonium salts as visualisation reagents for the thin layer chromatographic characterisation of amphetamines', *Sci. Justice*, **35**, 37–44 (1995).
7. Moffat, A. C., Jackson, J. V., Moss, M. S., Widdop, B. and Greenfield, E. S., *Clarke's Isolation and Identification of Drugs in Pharmaceuticals, Body Fluids and Post-Mortem Material*, Pharmaceutical Press, London, 1986.
8. Lillsunde, P. and Korte, T., 'Determination of ting and *N*-substituted amphetamines as heptafluorobutyryl derivatives', *Forensic Sci. Int.*, **49**, 205–213 (1991).
9. Esseiva, P., Lock, E., Gueniat, O. and Cole, M. D. 'Identification and quantification of amphetamine and analogues by capillary zone electrophoresis', *Sci. Justice*, **37**, 113–119 (1997).
10. Lurie, I. S., Bethea, M. J., McKibben, T. D., Hays, P. A., Pellegrini, P., Sahai, R., Garcia, A. D. and Weinberger, R., 'Use of dynamically coated capillaries for the routine analysis of methamphetamine, amphetamine, MDA, MDMA, MDEA and cocaine using capillary electrophoresis', *J. Forensic Sc.*, **46**, 1025–1032 (2001).
11. White, P. C., personal communication.
12. Jonson, C. S. L., personal communication.
13. Jonson, C. S. L. and Stromberg, L., 'Computer aided retrieval of common batch members in Leuckart amphetamine profiling', *J. Forensic Sci.*, **38**, 1472–1477 (1993).
14. Jonson, C. S. L. and Stromberg, L., 'Two-level classification of Leuckart amphetamine', *Forensic Sci. Int.*, **69**, 31–44 (1994).
15. Jonson, C. S. L. and Artizzu, N., 'Factors influencing the extraction of impurities from Leuckart amphetamine', *Forensic Sci. Int.*, **93**, 99–116 (1998).
16. Rashed, A. M., Anderson, R. A. and King, L. A., 'Solid-phase extraction for profiling of ecstasy tablets', *J. Forensic Sci.*, **45**, 413–417 (2000).
17. Ballany, J., Caddy, B., Cole, M., Finnon, Y., Aalberg, L., Janhunen, K., Sippola, E., Andersson, K., Bertler, C., Dahlén, J., Kopp, I., Dujourdy, L., Lock, E., Margot, P., Huizer, H., Poortman, A., Kaa, E. and Lopes, A., 'Development of a harmonised pan-European method for the profiling of amphetamines', *Sci. Justice*, **41**, 193–196 (2001).

Chapter 3

The Analysis of LSD

Learning Objectives

- To have an appreciation of LSD as a controlled substance.
- To be aware of the sampling procedures and descriptions required for LSD analysis.
- To understand the extraction of LSD from blotter acids for analysis.
- To be aware of the chemical and fluorescence testing procedures for LSD.
- To understand the principles of TLC analysis of LSD.
- To be aware of the confirmatory techniques available for the analysis of LSD.

3.1 Introduction

While a large number of drugs are known which are of plant origin, or have plant products as starting materials for the synthesis of the drugs, there are, equally, a number of drugs of fungal origin. Of these, perhaps lysergic acid diethylamide (LSD) (1) is the most 'famous', i.e. well known, and it is on this drug that this chapter focuses. Interestingly, the drug is an indole alkaloid which presents special difficulties and opportunities in terms of drugs analysis.

Lysergic acid diethylamide (LSD) is one of the most potent hallucinogens known to man. It was first synthesized in 1938 and was discovered to be psychoactive in 1943. It was initially used, experimentally, in the treatment of mental disorders but has not been used in this way for some 30 years. LSD encountered in the illicit drugs market of today is produced in clandestine laboratories. These are rarely detected because they make a large quantity of LSD, which lasts for an extremely long period of time, since only very small doses are administered and subsequent syntheses are not required [1]. LSD is, in the main, prepared from

$$CON(C_2H_5)_2$$

1

lysergic acid, via a series of complex reactions which require careful monitoring and control. The forensic scientist will see the resulting drug in a number
of differing dosage forms. The materials may be added to inert substrates or to
sugar cubes, or mixed with molten gelatin which is then cooled and cut into
small pieces containing the appropriate dose. These latter are known as 'window
panes'. However, these dose forms suffer from great inhomogeneity and the vast
majority of LSD observed in the forensic science laboratory today is encountered
in the form of 'blotter acid'. In this form, an absorbant paper is dipped into a
solution of LSD, and then dried. Such a procedure allows an even distribution
of the drug through the paper. A typical blotter acid dose contains between 30
and 150 μg of LSD per dose. Blotter papers are frequently decorated, with some
examples being shown on Plates 3.1 and 3.2, and represent the dose form on
which our discussion will centre.

   LSD doses vary between users and with the desired effect. Commonly, the
doses lie in the range between 30 and 150 μg of LSD per dose, although stronger
doses of 150 to 400 μg of LSD per dose (or even higher) are sometimes encountered. The onset of effect takes between 30 and 120 min and lasts between 6 and
14 h. After-effects can last for up to 24 h.

   In terms of legislative control, in the United Kingdom lysergamide, lysergide
and any *N*-alkyl derivatives of lysergamide are controlled as Class A drugs. In
the United States, LSD is controlled as a Schedule I drug. As part of the forensic
science process, therefore, it is necessary to prove the presence of the drug in
any sample thought to contain LSD.

3.2 Qualitative Identification of LSD

The process of analysis of LSD blotter acid follows the same general principles
and sequence as for other controlled substances, namely physical description,
presumptive testing, TLC and confirmatory analysis. These processes are discussed below.

Plate 3.1 Bird of paradise pattern on LSD blotter acid, covering the whole sheet of the paper. Copyright Michael D. Cole, Anglia Polytechnic University, Cambridge, UK, and reproduced with permission.

3.2.1 Sampling and Physical Description of LSD Blotter Acid

As with all forensic science analyses, the first stage in the process is a full physical description of the material under investigation. In the case of blotter acids, this includes a count of the number of dose units, the size of each of the dose units (length × breadth), whether they fit together, the number and depth of

Plate 3.2 Illustration of a 'ghost' on LSD blotter acid, with each image covering a few dose units. Copyright Michael D. Cole, Anglia Polytechnic University, Cambridge, UK, and reproduced with permission.

the perforations of the dose units and a note of the pattern and whether it covers all, some, or single dose units (see Plates 3.1 and 3.2).

DQ 3.1

Why is it necessary to record in such detail the physical information relating to an LSD seizure?

Answer

By recording such information, it may be possible to relate one or more samples to each other. This may be especially important when seizures from different occasions are being compared, or comparison between seizures is being made in different laboratories.

Having carried out a full physical description of the seizure, items must be chosen for analysis. While it might be assumed that the dose units are all identical, this might not be the case and it is therefore necessary to sample a number of them from the seizure. The United Nations Drug Control Programme (UNDCP) recommends the following procedure. For sample sizes up to 10 dose units, all should be analysed, while for sample sizes between 11 and 27 dose units, three quarters of the items should be selected at random (the number being rounded up to the next highest integer). For sample sizes in excess of 28 dose units, 50% should be selected at random, with a minimum of 21 dose units up to a maximum of 50 units [2].

DQ 3.2

How might random samples be chosen?

Answer

A random sample means that every item in a population has an equal chance of being chosen. Simply choosing materials by eye does not satisfy this criterion. Each of the dose units should be assigned a number, starting at 1 and ending with the last number (i.e. the number of items in the sample). The materials to be chosen should then be picked by using either a computerized random-number generator or random-number tables. Whichever method is used, it should be documented.

3.2.2 Extraction of LSD Prior to Analysis

Since the drug is impregnated onto a paper substrate, it is necessary to extract the material prior to analysis. In order to do this for presumptive testing or qualitative analysis, the extraction can simply be achieved by mixing the test sample for 30 s with sufficient methanol to achieve a sample concentration of 1 mg LSD ml^{-1} [2]. Alternatively, a methanol/water (1:1) mixture has been reported to extract the LSD more efficiently [3, 4]. It should be remembered that any solid material should be removed from the extract prior to any chromatographic analysis being carried out. This can be achieved either by centrifugation or by passing the extract through a 5 μm filter.

SAQ 3.1

Why should the solid material be removed prior to analysis?

If quantitative analyses are to be carried out, it is necessary to completely extract the LSD from the paper. This can be achieved by suspending the material in a large volume (15 ml is suggested) of 1% tartaric acid solution in a separatory funnel. The mixture is extracted three times with an equal volume of chloroform and then the aqueous layer is basified with 1 M $NaHCO_3$. The resulting mixture should be extracted, three times, with an equal volume of chloroform, and the chloroform extracts combined, filtered or centrifuged, and evaporated under a stream of nitrogen. The residue should then be reconstituted in a known volume of solvent [2]. Other acids have been used (summarized in Veress [3]), although the physico-chemical principles remain the same.

SAQ 3.2

What are the physico-chemical principles behind using an acid to extract the LSD from the paper and the subsequent processing?

3.2.3 Presumptive Testing for LSD

The presumptive tests for LSD involve a fluorescence test and a chemical test.

3.2.3.1 Fluorescence Testing for LSD

One of the properties of LSD that can be exploited during the identification process is its fluorescence. In such a test, the original dosage form, or a drop of the methanolic extract from the dose form, is placed on a filter paper and allowed to dry. The material is then observed under long-wavelength UV light (360 nm). If LSD is present, blue fluorescence will be observed.

DQ 3.3

If a methanol extract solution is used, what are the appropriate positive and negative controls and what do they show?

Answer

If methanol is used, the appropriate negative control is methanol alone (not exposed to sample or drug). The relevant positive control is a methanolic solution of LSD at approximately the same concentration as the sample provides.

The negative control demonstrates that the fluorescence, if observed, is due to the drug extracted into the methanol. The positive control provides a reference colour reaction and gives an idea of the intensity of the fluorescence that might be observed.

3.2.3.2 Chemical Testing for LSD

The classical presumptive test for LSD is the Ehrlich's reagent test. In this, 1 g of the reagent (*p*-dimethylaminobenzaldehyde) is dissolved in 10 ml of orthophosphoric acid. A small amount of the reagent solution is then added either directly to the test substrate or to the extract to be tested and any colour change observed.

If a blue/purple colour develops, then the presence of LSD may be suspected and confirmatory tests should be carried out.

DQ 3.4

Why is a confirmatory test required?

Answer

Ehrlich's reagent reacts with a wide range of controlled substances and other indole alkaloids. A positive reaction does not therefore prove the presence of a 'specific' drug. This is why it is necessary to carry out an additional confirmatory test.

3.2.4 Thin Layer Chromatography of Samples Containing LSD

Following the possible identification of the presence of LSD, the next stage in the analysis is the use of thin layer chromatography (TLC). This is employed because although it cannot be used to *prove* the identity of LSD, it can be used as a rapid, cost-effective method to eliminate those samples which gave a positive colour reaction in the presumptive tests but which do not contain this drug. These will be rarer when blotter acids are suspected, but may be more common where other substrates have been used as the carrier medium for the LSD itself.

The extracts are prepared as described above (see Section 3.2.2). Activated silica gel plates containing a fluorescent dye (which fluoresces at 254 nm) are used. The materials to be tested, plus positive and negative controls, are spotted onto the plate and the chromatograms developed in the normal way. The solvent systems which can be used include chloroform/methanol (9:1, by volume) and chloroform/acetone (1:4, by volume). Following chromatogram development, the plates are removed from the chromatographic tank, the solvent fronts marked, and the plates air-dried and observed under short- (254 nm) and long- (360 nm) wavelength UV light. Under the former lighting conditions, LSD will appear as a dark spot on a light background, while under the latter conditions it will appear as a bright spot on a dark background. The chromatogram should then be developed with Ehrlich's reagent, with which indole alkaloids (including LSD) react to give a purple product. If a product gives the same results (retardation factor, R_f, and colour reaction) as LSD under all of the conditions described, then further confirmatory tests should be carried out. However, if the materials do not yield the same physico-chemical responses as LSD under all conditions, then an exclusion has been achieved.

3.2.5 Confirmatory Tests for the Presence of LSD

Since there are a large number of materials which will produce the same result as LSD in one or more of the screening tests, it is necessary to confirm the presence of this drug by using an instrumental technique. Of the methods available,

HPLC with fluorescence detection is the most commonly used, although GC techniques have been reported [4, 5]. HPLC is also used as the means of quantifying LSD.

3.2.5.1 HPLC Analysis of LSD

A number of HPLC methods are available for the analysis of LSD. Two are presented here, namely (i) an ion-suppression technique, and (ii) an ion-paired method. The majority of techniques that are used exploit the same physico-chemical characteristics of the systems as those described here.

The first method is a reversed-phase method, using the HPLC system conditions described in Table 3.1. LSD is particularly amenable to fluorescence detection, which also has the advantage of providing both selectivity and sensitivity.

SAQ 3.3

How is selectivity of detection achieved when using a fluorescence detector to detect analytes in an HPLC eluant?

DQ 3.5

How does the ion-suppression work in this system?

Answer

At pH 8.0, the LSD will be present in the free base form. It will not be highly charged and hence will exhibit good chromatographic properties.

The second of the methods [3] uses an ion-pairing system, the conditions of which are given in Table 3.2. This system is not as specific as that employing fluorescence detection, although it is included here to illustrate that other separation mechanisms can be employed.

SAQ 3.4

Why is this detection system not as specific as the fluorescence detection one?

Table 3.1 Typical HPLC (ion-suppression) operating conditions and parameters used for the identification and quantification of LSD [2]

System/parameter	Description/conditions
Column	ODS^a silica: 10 cm $\times$ 4.6 mm i.d.; 5 μm particle size
Solvent	Methanol (65%): 25 mM Na$_2$HPO$_4$, pH 8.0 (35%)
Flow rate	1 ml min^{-1}
Detection	Fluorescence: excitation λ, 320 nm; emission λ, 400 nm

^{a}ODS, octadecasilyl.

Table 3.2 Typical HPLC (ion-pairing) operating conditions and parameters used for the identification and quantification of LSD [3]

System/parameter	Description/conditions
Column	ODS^a silica: 25 cm × 4.6 mm i.d.; 0.8 μm particle size
Solvent	Acetonitrile (45%): 50 mM KH_2PO_4/5 mM $C_8H_{17}SO_3Na$, pH 3.5 (55%)
Flow rate	0.9 ml min^{-1}
Detection wavelength	220 nm

^{a}ODS, octadecasilyl.

SAQ 3.5

What is the chromatographic mechanism operating in the solvent system described in Table 3.2?

3.2.5.2 GC Analysis of LSD

Gas chromatographic analysis of LSD is problematic because of the relatively limited volatility of this (controlled) substance [3]. However, GC-based methods have been applied, e.g. GC–MS [5]. The conditions reported for such analysis are shown in Table 3.3.

SAQ 3.6

What are the potential difficulties in analysing LSD by using isothermal gas chromatography?

3.2.5.3 Identification of LSD by Microscope FTIR Spectroscopy†

More recently, it has become possible to identify LSD *in situ*, without the need for extraction of the drug [6]. In this technique, LSD blotter acid is 'described' (see

Table 3.3 GC–MS operating conditions and parameters used for the identification of LSD [5]

System/parameter	Description/conditions
Column	BP-5: 25 m × 0.2 mm i.d.; d_f, 0.33 μm
Injection temperature	270°C
Column oven temperature programme	100°C; rising at 24°C min^{-1} to 270°C; isothermal for 35 min
Detection	Mass spectrometric
Split ratio	— ^{a}

^{a}Not reported.

---

† This is a technique used to measure the IR adsorption or reflection spectra of very small samples. In this method, a sample is placed on a KBr disc and a microscope is then used to focus the IR beam onto the material.

Section 3.2.1) and then subjected to a simple extraction procedure, followed by microscope FTIR spectroscopy directly on the extract. In this approach, the blotter acid is first soaked in water for 1 s, which swells the fibres of the papers and was found to facilitate the extraction of the drug for further spectroscopic analysis. The excess water was then removed from the blotter, which was subsequently placed on a KBr disc and heated to 120°C for 1 min. Dichloromethane/ammonia (100:1) was added to the blotter, which dissolved the LSD. The solution was then removed by using a microsyringe, placed on a KBr disc, and its spectrum recorded.

DQ 3.6

Why is ammonia added to the dichloromethane?

Answer

LSD is a basic drug. The ammonia will convert any LSD from the salt form to the free base state. In this form, the LSD is more soluble in organic solvents such as dichloromethane and thus the extraction process will be more efficient.

The IR spectrum was obtained by scanning in the wavenumber range 700–4000 cm^{-1}, with 30 scans at a resolution of 4 cm^{-1} being collected. From such data it was possible to identify LSD directly.

SAQ 3.7

Why were 30 scans taken and what is the advantage of using the scan range from 700 to 4000 cm^{-1}?

Summary

LSD is one of the most potent hallucinogens known to man. This drug can occur in the form of either tiny tablets or as gelatin blocks, but is most commonly found as 'blotter paper' – small pieces of highly decorated paper into which the LSD has been impregnated. The analysis of blotter acids includes a physical description of the item, followed by drug extraction from the paper. The extract is then used for fluorescence and colour tests to determine if LSD may be present. Following this, TLC and HPLC with fluorescence detection can be used to confirm the presence of the LSD, although GC methods are available for such analysis. An interesting new technique, using microscope FTIR spectroscopic analysis of the drug *in situ* on the paper, has also been developed.

Plate 3.1 Bird of paradise pattern on LSD blotter acid, covering the whole sheet of the paper. Copyright Michael D. Cole, Anglia Polytechnic University, Cambridge, UK, and reproduced with permission.

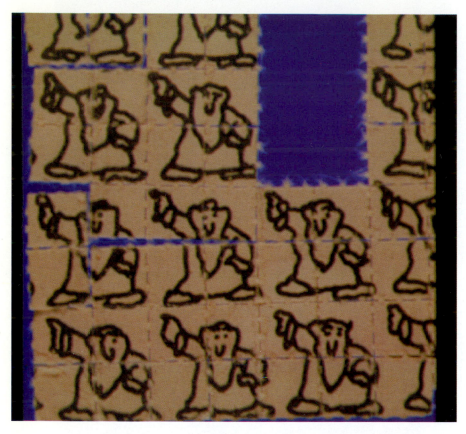

Plate 3.2 Illustration of a 'ghost' on LSD blotter acid, with each image covering a few dose units. Copyright Michael D. Cole, Anglia Polytechnic University, Cambridge, UK, and reproduced with permission.

References

1. Huizer, H., personal communication.
2. United Nations Drug Control Programme, *Recommended Methods for Testing Lysergide (LSD)*, Manual for use by national narcotics laboratories, United Nations Division of Narcotic Drugs, New York, 1989.
3. Veress, T., 'Study of the extraction of LSD from illicit blotters for HPLC determination', *J. Forensic Sci.*, **38**, 1105–1110 (1993).
4. Ripani, L., Schiavone, S. and Garofano, L., 'GC Quantitative determination of illicit LSD', *J. Forensic Sci.*, **39**, 512–517 (1994).
5. Kilmer, S. D., 'The isolation and identification of lysergic acid diethylamide (LSD) from sugar cubes and a liquid substrate', *J. Forensic Sci.*, **39**, 860–862 (1994).
6. Hida, M. and Mitsui, T., 'Rapid identification of lysergic acid diethylamide in blotter paper by microscope FT-IR', *Anal. Sci.*, **15**, 289–291 (1999).

Chapter 4

Cannabis sativa and Products

Learning Objectives

- To be aware of the origins and the manufacturing processes of different cannabis products.
- To understand how cannabis and its various products can be identified.
- To learn how cannabinoids can be quantified.
- To appreciate how cannabis samples can be compared.
- To be aware of the correct interpretation following the analysis of cannabis samples.
- To appreciate the difficulties associated with these methods.

4.1 Introduction

The history of the use and identification of *Cannabis sativa* L. is long and complex. It is one of the oldest cultivated plants, used for the production of oil from the seeds, and fibre from the stems for rope and fabrics, and has long been used as a psychoactive drug due to the presence of cannabinoids in the resins produced by the plant. Indeed, there is evidence of cannabis use from Neolithic burial sites. The name *Cannabis sativa* was first used in Linnaeus' *Genera Plantarum* in 1753, but since that publication there has been considerable debate about the number of species and varieties that exist – this has been recently summarized succinctly by Gigliano [1]. The debate has centred on (i) the characteristics of the fruit, and (ii) meiosis and pollen fertility. It is now generally accepted, however, that there is only one species, namely *Cannabis sativa* L., which exhibits great diversity due to both selection in the wild and in the cultivated environment.

Of the drugs that are contained in cannabis products, it is Δ^9-tetrahydrocannab-inol (Δ^9-THC) (**1**) which is responsible for the pharmacological activity of cannabis. This compound is formed in the glandular trichomes (see below) which are found on the surface of the plant. *Cannabis sativa* is dioecious, that is, it has both male plants and female plants. These are most easily recognized at the flowering stage because the flower structures are different. The female plants are preferred because they produce more of the glandular trichomes and, as a consequence, are richer in cannabinoids.

1

Within the United Kingdom, cannabis is controlled under the Misuse of Drugs Act, 1971, as amended by Section 52 of the Criminal Law Act, 1977, where controlled cannabis plant material is defined as follows:

any plant of the genus Cannabis *or any part of such plant (by whatever name designated) except that it does not include cannabis resin or any of the follow-ing products after separation from the rest of the plant, namely:*

(a) mature stalk of any such plant;

(b) fibre produced from mature stalk of any such plant;

(c) seed of any such plant.

The legislation is further complicated because in Europe, cannabis used for the manufacture of cooking oils is also recognized. While possession of the seeds is not in itself illegal, it is illegal to plant them and grow cannabis from the seeds.

At the time of writing,[†] herbal cannabis products are considered Class B drugs under the Misuse of Drugs Act, 1971, although discussion is currently under-way whereby the material will be reclassified. Cannabis resin, obtained from collecting the materials on the surface of the plant, is also a Class B drug, but again discussion is underway for rescheduling the materials. The cannabi-noids, namely cannabinol itself and cannabinol derivatives, including the active

---

[†] May, 2002.

ingredient Δ^9-tetrahydrocannabinol, are included as Class A drugs since they are considerably more potent than the herbal or resinous material. Cannabinol derivatives are defined in Part IV of Schedule 2 of the Misuse of Drugs Act, 1971 to mean the following:

... the following substances, except where contained in cannabis or cannabis resin, namely tetrahydro derivatives of cannabinol and 3-alkyl homologues of cannabinol or one of its derivatives.

The final type of cannabis product that is likely to be encountered is cannabis oil, more commonly known as 'hash oil' which is obtained by solvent extraction of the herbal or resinous material. There is debate in the legal literature as to whether this constitutes a Class A or Class B drug, the key to which appears to be the presence of cannabidiol. If the latter is present, then the material is treated as a purified form of resin (Class B), while if it is absent, the material is considered to have been *prepared*, and therefore falls under Class A.

4.2 Origins, Sources and Manufacture of Cannabis

The plant material can be used as herbal material, once dried, e.g. marijuana. Low-quality products, which contain stalks, seeds, leaves and flowering tops, may be compressed into blocks (West African and Caribbean material), it may occur as loose herbal material (from Central and Southern Africa), or it may be rolled into a so-called 'Corn Bob', wrapped in vegetable fibre (again from Central and Southern Africa). Higher-quality materials, composed of fruiting tops and flowers alone, may also be encountered. If tied around bamboo sticks, this material is known as 'Buddha Sticks' or 'Thai Sticks', and arises from South-East Asia. A central bamboo cane is used, around which up to 2 g of herbal material can be tied. The materials can be seized in bundles of up to 20 sticks. An African equivalent is to wrap the material in a small roll of brown paper; such rolls frequently contain less than 0.5 g of cannabis per roll. Sieved products may also be encountered. This process removes the stems and the leaves, producing 'Kif', a material derived from North Africa, for example, from Morocco.

Alternatively, the resinous material can be collected from the surface of the plant, dried and pressed into blocks. Resin production is centred in two parts of the world, i.e. the Southern and Eastern parts of the Mediterranean, and the Indian Subcontinent. This has essentially led to two types of cannabis resin.

Material from the Mediterranean region is prepared by threshing the herbal material, often against a wall, to separate the resin-producing parts from the non-producing parts, so detaching the resin, seed and leaves. The material is then sieved to remove the seeds and any stem material that are present, and is finally compressed into slabs for distribution. It may be wrapped in cloth bags, or in cellulose wrapping, prior to compression.

Resin from the Indian Subcontinent is prepared in a different way. Plants from this region are particularly sticky and resinous, and the resin transfers from the plant surface to any surface onto which the plant is rubbed. The resin is thus obtained by rubbing the plants against the palms of the hands, or against rubber sheeting. The resin is then collected, and shaped into slabs, rods or balls, or any other desired shape.

Finally, the cannabinoids can be extracted from the herbal material or the resin to produce hash oil. The latter is obtained from the extraction of cannabis plant material with a suitable organic solvent (for example, petrol or ether) by refluxing. Once a solution of the required strength has been obtained, the solvent is evaporated and the oil concentrated, giving a sticky green, olive or brown residue.

Herbal leaf material contains approximately 1% Δ^9-THC, while flowering material contains 3.5% of the active component [2], although there is a trend towards plant material of increasing potency in some parts of the world, with material containing up to 4.2% Δ^9-THC. Resin material contains 2–10 wt% Δ^9-THC, and hash oil 10–30 wt%.

Cannabis products are administered in a number of different ways. The most common method involves mixing with tobacco and smoking. A typical hand-rolled 'joint' contains approximately 100 mg of resin or 260 mg of herbal material [3], although the amount used will depend upon the strength of the material and the requirements of the user. Hash oil can also be smoked, by dropping the oil onto tobacco prior to rolling the cigarette. In addition, special pipes and paraphernalia are also associated with smoking cannabis. Alternatively, a drink ('Bhang') may be made from the material, or special foodstuffs can be prepared into which cannabis products have been incorporated. It should be noted that the most efficient method of ingestion also involves heating the cannabis product to above 100°C. This is because the process of heating causes Δ^9-tetrahydrocannabinolic acid (**2**), which is also found in cannabis products, to thermally decarboxylate to Δ^9-THC, thus increasing the potency of the material [4].

2

4.3 Analytical Sequence, Bulk and Trace Sampling Procedures

Prior to the process of identifying whether or not a material contains cannabis or its products, a decision must be made as to whether the sample is a bulk or a trace sample. A bulk sample can be defined as anything that can be seen (herbal material, resin or oil), while a trace sample, in terms of cannabis, can be loosely defined as one which would easily be contaminated. Trace samples might be present, for example, on the surfaces of scales used to weigh cannabis materials, the surfaces of knives used to cut resinous blocks or the surfaces of paraphernalia associated with cannabis use. Bulk samples range from a few mg of plant, resin or hash oil, to kilogram and tonne seizures arising from intervention of law enforcement agencies in smuggling operations. Their analysis follows the sequence, physical description, presumptive test, thin layer chromatography and then instrumental analysis.

The analytical sequence for trace samples involves a physical description of the item, followed by sample collection and instrumental analysis directly. Trace samples are analysed while using all of the precautions required to prevent contamination of materials. The laboratory, equipment, chemicals and operators must be demonstrably 'free of cannabis' and should be tested prior to sampling of the casework materials. The latter should then have samples taken from them. This is usually undertaken by swabbing part of the surface to be considered with a cotton wool swab soaked in ethanol. The latter is used because the cannabinoids have been demonstrated to be stable in this solvent and to be freely soluble. While the cannabinoids are stable in diethyl ether, the more polar acids are not freely soluble in this solvent. While chloroform is a solvent in which the cannabinoids are freely soluble, it should not be used because it contains HCl which is known to catalyse the breakdown of the cannabinoids.

SAQ 4.1

Why should the whole surface of a trace item not be swabbed?

Bulk samples come in a large variety of sizes. If they are wrapped, consideration should be given to preservation of the wrapping because this might allow (i) links between samples to be established, and (ii) fingerprints to be recovered. The materials should be taken from the bulk of the drug, that is, *not* the outer layer. This can be achieved by using a cork borer to bore into the herbal or resinous material in order to collect the sample for analysis. It should be remembered that this should never be carried out on a surface on which a physical fit with another item might be established, or where there are surface markings or striations which might prove to be of significance.

A good physical description of the items, in particular, bulk samples of cannabis, should include the size, weight, colour, smell, shape and physical relationship of each of the samples.

DQ 4.1

What is the value of weighing the material prior to its analysis?

Answer

This process will allow a decision to be made as to whether the material is for personal use, or is to be supplied to others. For example, samples of resin are frequently encountered in 250 g 'soap bars', which represents a considerable amount of material. This is then sub-divided into smaller amounts, and finally into single doses of between 100 and 400 mg, e.g. for each cigarette. Knowing the average weight of material contained within a cigarette and the amount of seized material present, calculations can be made as to the number of cigarettes that can be potentially made from the material in the sample.

DQ 4.2

What is the value of assessing the physical relationship of different pieces of cannabis resin?

Answer

If two resin block fit together, this proves that they were once part of a single, larger piece. No further chemical analysis is required to establish this. If, however, they do not fit together and it is still required that it be determined whether or not they are related, then further chemical analysis is required.

4.4 Qualitative Identification of Cannabis

The methods used to identify cannabis products depend upon the nature of the products themselves. Herbal material can be identified on the basis of its morphological characteristics alone, provided that certain of these are present. Where they are not, and in the case of resin and hash oil, the identification is made on the basis of phytochemical identification and the proof of the presence of Δ^9-THC (**1**), its precursor, cannabidiol (CBD) (**3**) and its breakdown product, cannabinol (CBN) (**4**). However, it should be remembered that the presence of the breakdown product, CBN, precludes the use of the sample for comparative purposes.

4.4.1 Identification of Herbal Material

Botanically, *Cannabis sativa* can be identified on the basis of its gross morpho-logical features and, more importantly, by the presence of microscopic structures on the surface of the plant, namely the trichomes. At the macro-morphological level, it has a square stem, with four corners, and has palmate leaves with ser-rated edges. These are the characteristics with which most people are familiar. Microscopically, three types of trichome are observed, namely the glandular tri-chomes (Plate 4.1), unicellular trichomes (Plate 4.2) and cystolithic trichomes (Plate 4.3).

Plate 4.1 Glandular trichromes found on the surface of *Cannabis sativa*. Copyright Michael D. Cole, Anglia Polytechnic University, Cambridge, UK, and reproduced with permission.

Plate 4.2 Unicellular trichromes found on the surface of *Cannabis sativa*. Copyright Michael D. Cole, Anglia Polytechnic University, Cambridge, UK, and reproduced with permission.

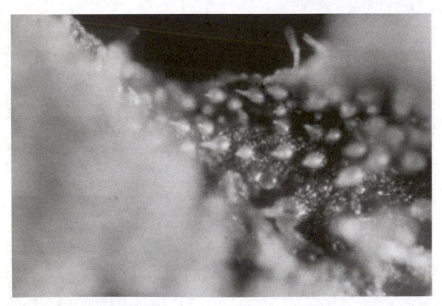

Plate 4.3 Cystolithic trichromes found on the surface of *Cannabis sativa*. Copyright Michael D. Cole, Anglia Polytechnic University, Cambridge, UK, and reproduced with permission.

The glandular trichomes are the structures in which the cannabis resin is produced. These can be found on the underside of the leaves, and occasionally on the stems, but are mainly associated with the flower structures, with female plants being particularly rich in such structures. The unicellular trichomes are found all over the plant, but interestingly, only 'point up' the plant. The cystolithic trichomes are found all over the plant and may have oxalic acid crystals visible within their bases. If all of these trichomes are found together, then the plant material can be definitively identified as *Cannabis sativa*, since no other plant displays such a combination [5]. However, some plants possess trichomes which may be confused with those present on *Cannabis sativa* [6, 7] and care should thus be taken in definitive identification.

4.4.2 Identification of Other Materials

If the morphological characteristics of *Cannabis sativa* are not all present, or if they have been destroyed, in, for example, resinous material or material that has been smoked (Plate 4.4), it is necessary to identify the material through its phytochemical content.

This is also true for hash oil, and all three materials will be considered in the following sections. This involves the identification of the principle drugs which characterize the plant, namely Δ^9-tetrahydrocannabinol (THC), CBD and CBN. For bulk samples, the analysis follows the sequence, presumptive test, TLC and

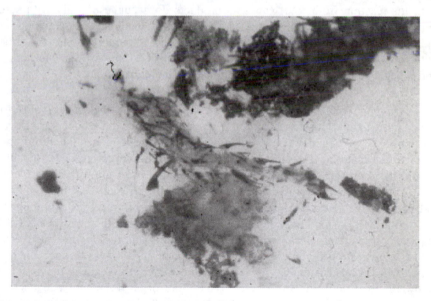

Plate 4.4 Unicellular trichromes present in the remains of smoked cannabis material. Copyright Michael D. Cole, Anglia Polytechnic University, Cambridge, UK, and reproduced with permission.

then identification, usually by GC–MS. Trace samples are analysed directly by also using GC–MS.

4.4.2.1 Presumptive Tests for Cannabis

Presumptive tests for cannabis products are used to test for the presence (or otherwise) of phenolic cannabinoids. The principle one that is used is the Duquenois–Levine test. An alternative which is available and can also be used is the Corinth IV salt test. Both involve a reaction between the cannabinoids and a test reagent to form coloured products. Positive and negative control tests should also be carried out.

The Duquenois–Levine Test In this, three reagents are required (see Appendix 1), with the reaction being carried out in small test-tubes. The three reagents are (i) 2.5 ml of acetaldehyde and 2 g of vanillin dissolved in 100 ml of 95% ethanol, (ii) concentrated hydrochloric acid, and (iii) chloroform. In the test-tube, five drops of the first reagent are added to the material to be tested which has previously had a drop of ethanol added to it in order to solubilize the cannabinoids. The mixture should be shaken, five drops of the HCl added, the mixture shaken again, 10 drops of $CHCl_3$ added, and the mixture shaken again. The two phases are then allowed to separate. Cannabinoids rapidly form a blue/purple complex which extracts into the chloroform layer. False positives that can be confused with the reaction from 'old' cannabis products do occur, especially with some herbs and some brands of instant drinks. However, the colour and extraction into the chloroform layer are not the same as when the coloured complex is formed with fresh cannabis material. Both a positive control (a known cannabis sample) and a negative control (ethanol) should also be undertaken.

The Corinth IV Salt Test This test relies on the extraction of the cannabinoids, followed by reaction with the specific test reagent. As with the Duquenois–Levine test, positive and negative controls should be performed for each batch of experiments. The material to be tested is placed on a filter paper and a second filter paper placed on top of this. A few drops of diethyl ether are placed on the top paper – above the material being tested – and the paper 'squashed' into the material, to 'mop-up' the ether extract. Next, the top paper is removed and the ether allowed to dry. A small amount of the 1% Corinth IV salt in anhydrous Na_2SO_4 is placed on the paper between the edge of the latter and the dried extract. Water is then gently dropped onto the outside of the paper, and moves through the filter paper; when it reaches the Corinth IV salt, it dissolves the reagent, thus carrying it towards the dried cannabis extract. When the Corinth IV solution reaches the cannabinoids, pink and orange products are formed, thus indicating the possible presence of cannabis. Again, positive (a known drug) and negative (ether only) controls should be undertaken.

DQ 4.3

Why are both a positive and negative control required?

Answer

The positive control is required to (i) show that the test is working, (ii) provide a reference colour against which the sample can be tested, and (iii) provide a reference time-frame in which the colour reaction might be expected to occur. These are required because some materials, for example, some brands of instant coffee, will give a so-called 'false positive' colour reaction, but much more slowly than a cannabis sample and the colour of the purple complex is not the same. The negative control is required to demonstrate that (i) the equipment is clean, and (ii) that the colour reaction is due to the presence of cannabis in the questioned sample.

Other presumptive tests are available for the tentative identification of cannabis [8] but these are not specific to this controlled substance because a wide range of plant materials also contain phenolic compounds which result in similar reactions to those obtained from cannabis. As a consequence, further methodologies must be used to identify the materials.

4.4.2.2 Thin Layer Chromatography of Cannabis

Thin layer chromatography provides a rapid, easy and cost-effective means for screening for cannabis materials and making comparisons between seizures. The cannabinoids are readily oxidized and so should be prepared for TLC in a solvent in which they are stable. As with the swabs for trace samples, ethanol satisfies this criterion at this stage.

The samples should be prepared by suspension in ethanol at a concentration which will yield a cannabinoid content roughly the same as for the positive controls. Having extracted the materials for a given period of time, the solid material should be removed. This is most easily achieved on samples of this size by using centrifugation. The supernatant is retained for subsequent analysis.

SAQ 4.2

Given the solubility of the cannabinoids and the carboxylic acids of cannabinoids, why is diethyl ether not a suitable solvent for cannabis products?

SAQ 4.3

Why does the acid lability of the cannabinoids preclude the use of chloroform as a suitable solvent for sample preparation of cannabis products prior to analysis?

The extracts, positive control (cannabinoids, at a known concentration) and negative control (the same batch of ethanol as the one used for the samples) are spotted onto a silica chromatographic plate and allowed to dry. The extracts and standards should contain between 5 and 10 µg of cannabinoids. The TLC mobile phase is placed in the tank and the system allowed to equilibrate. The most frequently used mobile phase is toluene, although other systems can be used [9, 10]. The chromatogram is developed, removed from the solvent and the mobile phase allowed to evaporate at room temperature. If toluene (or another dangerous solvent) is used, drying of the chromatographic plate should be carried out in a fume cupboard. Once the plate is dry, the compounds present should be examined visually under white and UV light (254 nm and 360 nm) and after spraying with an appropriate reagent. The most commonly used reagent is 1% Fast Blue BB in water, which has replaced the previous use of the carcinogenic Fast Blue B. A variety of red, orange, brown and yellow compounds form as a result of the reaction of the Fast Blue BB with the cannabinoids to form diazonium salts (for example, **5**).

5

While a number of plant phenolics will react with Fast Blue BB (which is why it cannot be taken as a method for the *definitive* identification of cannabis), it is the pattern of products which can be used to infer the material is cannabis and hence the necessity to proceed to further methods of proof. However, this method is extremely useful for making preliminary comparisons between samples. If the chromatograms of two further samples are different, the samples may not be related. If, however, the chromatograms cannot be differentiated, further investigation is required.

DQ 4.4

Why is it necessary to apply both a positive and a negative control to each thin layer chromatogram?

Answer

This is needed for a number of reasons. The ethanol (negative control) is applied to demonstrate that any compound which reacts with the visualization reagent has originated from either the standard solution or from the sample. The positive control (e.g. cannabinoids) is necessary to (i) demonstrate that the chromatographic procedure is working, (ii) show that the visualization reagent is working, (iii) provide reference R_f values, because these will be unique to each individual chromatogram, and (iv) provide reference colour changes.

SAQ 4.4

If the stationary phase is silica gel and the mobile phase toluene, what are the principle interactions between the analytes and the stationary phase that are likely to result in analyte retention?

DQ 4.5

What is an additional value of TLC in cannabis analysis?

Answer

*There are a number of reasons for carrying out TLC analysis of cannabis. This method can be used to identify the **pattern** of compounds typical of cannabis. It can also be used to determine **whether or not cannabinol is present in the sample**. Cannabinol is the breakdown product of Δ^9-tetrahydrocannabinol. If this is present, the sample is understood to have started to decompose and it should not be used for comparative purposes. It is for this reason that analysis for comparative purposes is not generally carried out more than three months after sample seizure.*

4.4.2.3 Gas Chromatography–Mass Spectrometry of Cannabis

In order to identify cannabis products, one of the most frequently used methods is gas chromatography–mass spectrometry (GC–MS). The identification relies on the presence of Δ^9-THC, CBD and CBN in the sample, and to a lesser extent, the presence of Δ^8-THC (the isomer of the active principle of cannabis) and Δ^9-tetrahydrocannabinolic acid. The following presents a method that has been found to work, although there are a number of similar methods reported in the literature [9]. The sample is prepared for GC–MS analysis as follows:

1. The bulk samples are triturated in ethanol, while trace samples on swabs are eluted by using the same solvent. Ethanol is chosen because it satisfies all of the requirements of a good solvent prior to further analysis. Standards are

prepared in ethanol, in the same concentration range, and an ethanol negative control is also retained.

2. The samples are centrifuged to remove any solid material which might block the syringe or the gas chromatographic system. The supernatant is used for subsequent analysis.

3. The samples, standards and negative control are placed in GC vials and 'blown down' under nitrogen. Nitrogen is used to prevent sample decomposition. The residue is usually present as an oil.

4. The derivatization reagent (100 μl of 10% N,O-bis(trimethylsilyl)acetamide (N,O-BSA) in n-hexane, containing 0.1 mg ml^{-1} of n-dodecane as the internal standard, when one is used) is added to the sample, the sample vial sealed, the vials shaken and the drugs allowed to derivatize at room temperature. The derivatization reaction is as shown in Scheme 4.1.

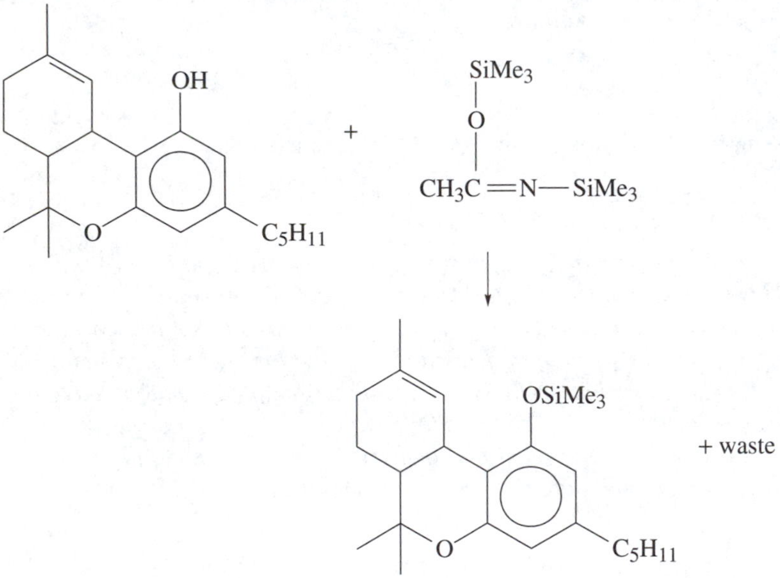

Scheme 4.1 Derivatization reaction of Δ^9-tetrahydrocannabinol with N,O-bis(trimethyl-silyl)acetamide.

5. The materials are analysed in the following order: derivatized blank, derivatized standard, derivatized blank, derivatized negative control, derivatized blank, derivatized sample, derivatized blank, etc., with a positive control at every 5th or 6th injection.

It should be noted at the outset that gas chromatography of the cannabinoids may present problems in that if derivatization is not employed, then the carboxylic acid compounds will decompose, since they are thermally labile.

DQ 4.6

Given the structures of the cannabinoids below, why is derivatization necessary?

OH

C_5H_{11}

Δ^9-tetrahydrocannabinol

OH

COOH

C_5H_{11}

Δ^9-tetrahydrocannabinol carboxylic acid

Answer

There are a number of reasons why derivatization of samples is necessary. First, the hydroxide and carboxylic moieties may form hydrogen-bonds in the gas phase, thus resulting in poor chromatography. Secondly, the compounds may sorb onto the chromatographic system, resulting in poor chromatography performance, peak shape, and in the case of very low concentrations, non-detection of the analytes. Furthermore, it is known that Δ^9-tetrahydrocannabinol carboxylic acid thermally decarboxylates at temperatures above 110–120°C, and converts into Δ^9-tetrahydrocannabinol. While in some laboratories this is assumed to be quantitative, this may not always be the case. Derivatization will allow determination of both the Δ^9-tetrahydrocannabinol and its carboxylic acid. In addition, in the case of very low concentrations of analytes, derivatization will improve the limits of determination and quantification.

Typical GC–MS operating conditions and parameters employed for the analysis of cannabis products are shown in Table 4.1.

A typical separation which can be achieved is shown in Figure 4.1, while the mass spectra of the trimethylsilyl (TMS) ethers of the derivatized cannabinoids are shown in Figures 4.2–4.5.

On the basis of the availability of the retention time data and the mass spectra of these compounds, and comparison of these between the compounds in the sample and the standards, an identification can be made. An example of a chromatogram obtained from herbal cannabis material (Figure 4.6) and the mass spectrum of the derivatized THC (Figure 4.7) are presented to illustrate this point.

Note that the relative intensities of the ions at m/z 386 and 371 are 'reversed' in the sample and standard. The intensities of these ions are dependant upon

Table 4.1 GC operating conditions and parameters suitable for the GC–MS analysis of cannabis products

System/parameter	Description/conditions
Column	OV-1 or equivalent: 25 m × 0.22 mm i.d.; d_f, 0.5 μm
Injection temperature	290°C
Column oven temperature programme	170°C for 2 min; increased to 280°C at 4°C min^{-1}; held for 2 min
Carrier gas	He, at a flow rate of 1 ml min^{-1}
Split ratio	50:1
Detector	Mass spectrometric, with temperature and settings as required

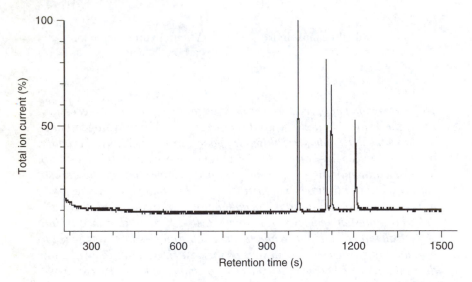

Figure 4.1 Total-ion chromatogram of separated trimethylsilyl ethers of the cannabinoids: cannabidiol (16.5 min); Δ^8-THC (18.3 min); Δ^9-THC (18.4 min); cannabinol (20.1 min).

a number of factors, including the amount of material reaching the detector. However, if the chromatographic peak is sectioned and the mass spectra compared to that of the standard, then the ions, in general, will be of the same mass-to-charge ratios and relative intensities as those obtained for the standard. This does, however, illustrate the point that even with definitive techniques such as GC–MS, there will be slight variations in the data between samples of which the forensic scientist should be aware.

SAQ 4.5

Why does mass spectroscopy allow the definitive identification of analytes?

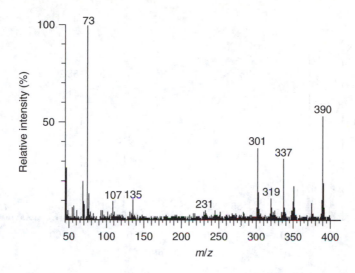

Figure 4.2 Electron-impact mass spectrum of the TMS ether of cannabidiol.

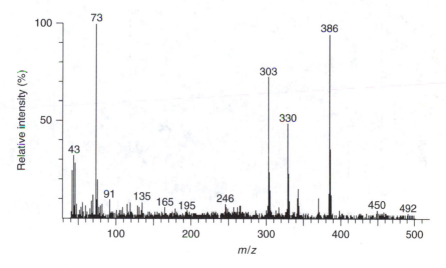

Figure 4.3 Electron-impact mass spectrum of the TMS ether of Δ^8-tetrahydrocannabinol.

4.4.3 *Comparison of Cannabis Samples*

On occasion, it may be necessary to make comparisons between cannabis samples. There are a number of questions which need to be addressed in order to establish or otherwise that two or more samples of cannabis once formed the same batch. These include the following:

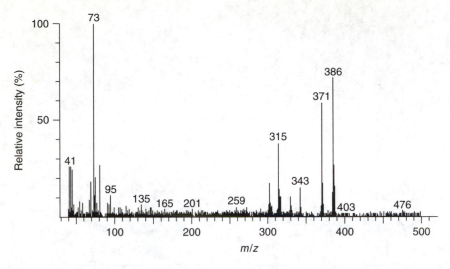

Figure 4.4 Electron-impact mass spectrum of the TMS ether of Δ^9-tetrahydrocannabinol.

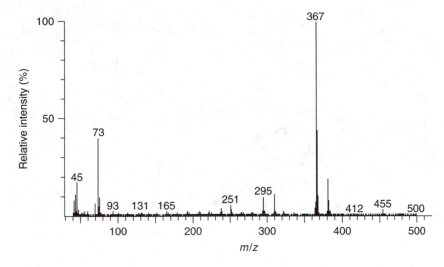

Figure 4.5 Electron-impact mass spectrum of the TMS ether of cannabinol.

1. Do the wrapping materials for the different samples fit together or are they physically or chemically related?

2. Do the drug samples fit together?

3. Are the drug samples chemically related?

4. Are the drug samples genetically related?

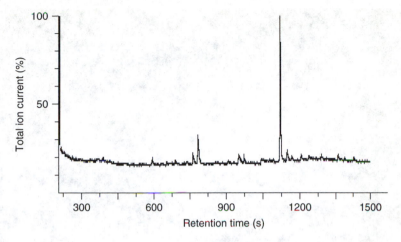

Figure 4.6 Gas chromatogram of cannabinoids from a herbal sample of cannabis after derivatization with N,O-BSA.

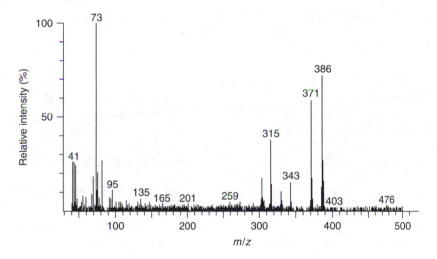

Figure 4.7 Mass spectrum of Δ^9-tetrahydrocannabinol from a herbal sample of cannabis after derivatization with N,O-BSA.

4.4.3.1 Physical and Chemical Comparison of Wrapping Materials

If there are more than one item in the exhibit/production label, each may be individually wrapped. The wrapping should be removed using very great care, since these materials themselves may fit together, thus indicating a relationship of the samples (Plate 4.5). This does not prove any chemical relationship between the samples, but they might have been wrapped by the same person, for example.

Plate 4.5 Wrapping materials, illustrating how a physical fit might be identified. Copyright Michael D. Cole, Anglia Polytechnic University, Cambridge, UK, and reproduced with permission.

Comparison of the striations on some extruded films, for example, cling film, may indicate whether or not the plastic film in which the drug samples are wrapped once formed a larger piece. Striations can be compared under both white and polarized light, and relationships established or otherwise. Paper wrappings are rather more problematic if they do not fit together, but under the microscope it may be possible to relate items together because the paper has the same microscopic appearance.

In addition to physical relationships, it might also be possible to establish chemical links between the wrapping materials. This is conveniently achieved by using Fourier-transform infrared (FTIR) spectroscopy. If the materials are chemically different, then an exclusion can be achieved. If they cannot be distinguished, further examination of the drugs is required.

Further to comparison between drug samples, it may also be possible to relate wrapping materials to someone or a place (a house, for example), or, indeed, to exclude such a relationship. However, whenever wrapping materials are compared, the drug materials should also be compared physically and chemically to determine whether there is a relationship or not.

4.4.3.2 Physical Comparison of Cannabis Samples

The most frequent comparisons of this type are between resinous samples. After examining the wrapping materials, the forensic scientist should make a physical

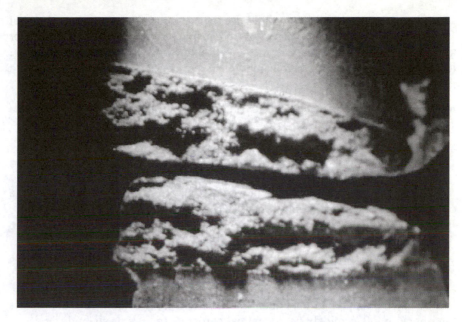

Plate 4.6 A low-power microscopic image of cannabis resin, illuminated to show raised areas on one common surface and the corresponding troughs on the opposite surface. Copyright Michael D. Cole, Anglia Polytechnic University, Cambridge, UK, and reproduced with permission.

description of the drug and then examine the pieces of the latter to determine whether or not they fit together (Plate 4.6). The examination should include examination of striae on the surface of the drug which might be used to link samples together, while stamps or marks made on the surface should also be compared. If the samples fit together, then an unequivocal relationship has been established and no further work is required to establish the link. However, if the samples do not fit together, then further chemical study is required.

4.4.3.3 Chemical Comparison of Cannabis Samples

Two principle methods are available for the analytical comparison of cannabis, namely high performance liquid chromatography and gas chromatography–mass spectrometry. Both techniques, along with their advantages and disadvantages are discussed below.

HPLC as a comparative technique has a number of advantages, including the fact that it does not require the analyte to be volatile, it does not require any pre-treatment of the sample prior to analysis, such as derivatization, it can be automated and can be used quantitatively. The HPLC technique used is a reversed-phase system and employs ion-suppression. Cannabinoids are weakly acidic drugs due to the presence of phenolic moieties on the aromatic rings.

Thus, in aqueous solution they will ionize, hence forming species which interact strongly with the stationary phase, and so causing tailing of the compounds. For this reason, acetic acid is added to the buffer in order to reduce the extent of ionization and improve the chromatographic performance. The system employed is as follows:

- Column: reversed-phase C-18 (ODS), 25 cm × 4.6 mm i.d.
- Eluant: methanol/water/acetic acid (85:14.2:0.8, by volume)
- Flow rate: 1.5 ml min^{-1}
- Detection: at 230 nm (using a diode-array detector)

The separation which can be achieved is illustrated in Figure 4.8. The data obtained illustrate the difficulty of HPLC comparisons, i.e. that of resolution. In this example, the two isomers of THC are not resolved with reference to the baseline.

When casework samples are being analysed, the problem of co-elution becomes even more evident (Figure 4.9). It is difficult to resolve all of the materials and diode-array spectra are not sufficiently discriminating to allow definitive identification. Coupled to long analysis times and short column life, this makes interpretation of seemingly near-identical chromatograms extremely difficult.

One way to overcome this problem is the use of GC–MS methodologies, as described above (Section 4.4.2.3). However, it should be remembered that for meaningful comparisons, the samples should be prepared in extracts of the same ratio of sample to solvent and complete derivatization should be ensured.

Nevertheless, even under such strictly controlled analytical conditions, there are difficulties associated with such approaches. If the breakdown product, cannabinol,

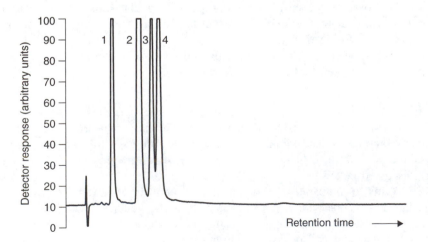

Figure 4.8 HPLC chromatogram of cannabinoids separated by using the HPLC system described in the text: 1, cannabidiol; 2, cannabinol; 3, Δ^9-THC; 4, Δ^8-THC.

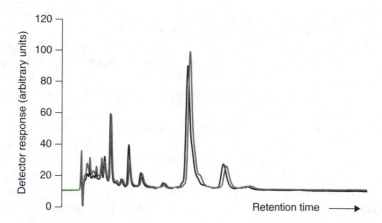

Figure 4.9 HPLC chromatograms of two casework samples of cannabis analysed under identical conditions to those of the standards.

is present and detected at the TLC stage, the material cannot be used for comparative purposes because of the differences in the rates of breakdown between the samples [9].

4.4.3.4 Comparison of Cannabis Using DNA Profiling

Recently emerging technologies, e.g. DNA profiling, have allowed many of the problems described above to be overcome. A discussion of the relevant DNA techniques is not required here. However, it is possible to (i) identify the presence of cannabis, and (ii) make comparison between samples by using such techniques [11–13, and references contained therein] and the drug chemist should be aware of the possibilities offered in this area. The reason that the DNA methodologies offer advantages over conventional chemical comparisons is that growers often cultivate plants by taking cuttings and these are genetically identical. Since cannabis is known to exhibit both within- and between-plant phenotypic plasticity, even when genetically identical plants are compared, these new approaches offer an advantageous way to determine where plant stocks have arisen.

Summary

Cannabis products can be found in a large number of forms, including herbal material, resin and oil. A wide variety of utensils associated with cannabis use may also be encountered by the forensic scientist. To determine whether or not cannabis products are present, a thorough physical examination of the material should be carried out. The next step is determined by the type of drug submitted for analysis. Herbal material may be identified directly by observation of the glandular, unicellular and cystolithic trichomes which are characteristic of this material. If these cannot be seen, then the material should be treated as for resin and oil. For these latter materials, physical descriptions are prepared, colour (presumptive) tests undertaken, and TLC analysis carried out, with confirmation of identity being obtained from GC–MS. HPLC should be used for sample comparison. Trace samples are analysed directly by GC–MS analysis of extracts of the alleged drug materials.

At the present time, HPLC is generally employed for sample comparison. However, recent developments in DNA methodologies have provided an alternative method of DNA profiling analysis for establishing relationships between samples. This is possible because maternally inherited DNA can be used to identify the relationships between plant cuttings. Such an approach overcomes the problems associated with phenotypic plasticity which has traditionally been used for sample comparison.

References

1. Gigliano, S. G., '*Cannabois sativa* L. – botanical problems and molecular approaches in forensic investigations', *Forensic Sci. Rev.*, **13**, 2–17 (2001).
2. Poulsen, H. A. and Sutherland, G. J., 'The potency of cannabis in New Zealand from 1976–1996', *Sci. Justice*, **40**, 171–176 (2000).
3. Buchanan, B. E. and O'Connell, D., 'Survey on cannabis resin and cannabis in unsmoked handrolled cigarettes seized in the Republic of Ireland', *Sci. Justice*, **38**, 221–224 (1998).
4. Moffat, A. C., 'The legalisation of cannabis for medical use', *Sci. Justice*, **42**, 55–57 (2002).
5. Segelman, A. B., Babcock, P. A. and Braun, B. L., '*Cannabis sativa* L. (marijuana) II: standardised and reliable microscopic method for the detection and identification of marijuana', *J. Pharm. Sci.*, **62**, 515–516 (1973).
6. Nakamura, G. R., 'Forensic aspects of of cystolith hairs of *Cannabis* and other plants', *J. Assoc. Off. Anal. Chem.*, **52**, 5–16 (1969).
7. Thornton, J. L. and Nakamura, G. R., 'The identification of marijuana', *J. Forensic Sci. Soc*, **12**, 461–519 (1972).
8. Tewari, S. N. and Sharma, J. D., 'Spot tests for cannabis materials', *Bull. Narcotics*, **31**, 109–112 (1982).
9. Gough, T. A. (Ed.), *The Analysis of Drugs of Abuse*, Wiley, Chichester, UK (1991).
10. Cole, M. D. and Caddy, B., *The Analysis of Drugs of Abuse: An Instruction Manual*, Ellis Horwood, Chichester, UK (1995).
11. Gillan, R., Cole, M. D., Linacre, A. M. T., Thorpe, J. W. and Watson, N. D., 'Comparison of *Cannabis sativa* by Random Amplification of Polymorphic DNA (RAPD) and HPLC of cannabinoids: a preliminary study', *Sci. Justice*, **35**, 169–177 (1995).
12. Jagadish, V., Robertson, J., and Gibbs, A., 'RAPD analysis distinguishes *Cannabis sativa* samples from different sources', *Forensic Sci. Int.*, **79**, 113–121 (1996).
13. Linacre, A. M. T. and Thorpe, J. W., 'Identification of cannabis by DNA', *Forensic Sci. Int.*, **91**, 71–76 (1998).

Chapter 5

Diamorphine and Heroin

Learning Objectives

- To be aware of the origins and manufacturing processes of heroin.
- To understand how heroin can be identified.
- To learn how heroin can be quantified.
- To appreciate how heroin samples can be compared.
- To be aware of the correct interpretation following the analysis of heroin samples.
- To appreciate the difficulties associated with these methods.

5.1 Introduction

In the United Kingdom, 'heroin', which has various street names, including 'dope', 'junk' and 'smack', is a generic term that describes a group of compounds which have been made during the preparation of diamorphine from the morphine extracted from the latex of the field poppy, *Papaver somniferum* L. In the United States, 'heroin' is used to mean diamorphine. On the street, the percentage of a heroin sample that can be diamorphine may range anywhere between 1 and 98 wt%, with average values lying between 35 and 41 wt%. The amount used in a dose depends upon the amount of diamorphine in the material and the tolerance and usage pattern of the individual using the drug. Those with little or no tolerance to the drug may administer between 5–15 mg by intravenous injection, 15–30 mg by smoking or 50–70 mg may be taken orally. Those with a much greater tolerance may administer as much as 20–60 mg by intravenous injection. In terms of efficiency, injection is the most effective way to administer the drug. However, with the fear of contracting syringe- and needle-borne diseases, including HIV and hepatitis, and the increase in the purity of the available heroin, in recent years there has been an increase in smoking and snorting of the drug. The effects induced include euphoria and drowsiness. However, the effects of overdose include slow and shallow breathing, clammy skin, nausea, convulsions, coma and possible death.

In the United Kingdom, diamorphine is controlled as a Class A drug, under Part 1 of Schedule 2 of the Misuse of Drugs Act, 1971, as is its precursor, morphine. Codeine is controlled as a Class B drug. In addition, the products found as impurities from diamorphine production, namely 3-monoacetylmorphine and 6-monoacetylmorphine, are controlled as esters of morphine, while acetylcodeine is controlled as an ester of codeine. In the United States, heroin (diamorphine) is controlled as a Schedule I narcotic.

5.2 Origins, Sources and Manufacture of Diamorphine

Heroin (diamorphine) is derived from morphine, which is obtained from the latex of the field poppy, *Papaver somniferum* L. As with all drugs, the sites of major production change with time. In 1995, it was estimated that the world opium harvest was 4157 tonnes, with the majority (2561 tonnes) being produced in South-East Asia (Myanmar (Burma), Thailand, China, Vietnam and Cambodia). More recently, Afghanistan has emerged as a leader of opium production, producing over 4500 tonnes in 1999, with the South-East Asian production dropping by about 50%. In addition, smaller-scale production occurs in Central and Southern America, with Colombia producing 60–70 tonnes of opium per annum, and Mexico producing 50 tonnes per annum. Heroin from these latter countries is destined for the United States, while that produced in South-East Asia is for global distribution, with that from South-West Asia (Afghanistan, Iran, Pakistan, India and Turkey) being destined for Europe or local, regional consumption.

Diamorphine is prepared by the isolation and acetylation of morphine. While morphine was first discovered in the dried latex of *P. somniferum* shortly after 1800, it is now known that of the approximately 110 species of *Papaver*, only two are known to produce morphine in significant quantities, namely *P. somniferum* L. and *P. setigerum* DC [1, 2]. Of the two species, it is *P. somniferum* which is used to provide the dried latex for heroin production.

Morphine is extracted from the dried latex (opium) from the seed capsule of the poppy. Incisions, about 1 mm deep, are made in the seed capsule approximately two weeks after the petals have fallen from the flower and the seed head is mature, and the latex exudes from the seed head, dries and oxidizes in the sun and is collected the following day. With successive harvestings, a pod may produce anywhere between 10 and 100 mg of opium. The latex contains 10–20 wt% alkaloids, with the remainder of the latex being sugars, proteins, lipids, gums and water.

Opium contains morphine (**1**) (4−21 wt%), codeine (**2**) (0.7−3 wt%) and the-baine (**3**) (0.2−1 wt%). In order that the morphine can be converted to diamor-phine (**4**) (the desired drug) it is necessary to (i) extract and purify the morphine from the opium, and (ii) synthesize and purify the diamorphine. As a consequence of co-extraction from the opium, heroin also contains noscapine (**5**) (2−8 wt%) and papaverine (**6**) (0.5−1.3 wt%).

In order to extract the morphine, the opium resin must first be prepared. This is achieved by adding the raw opium to boiling water, in which the alkaloids dissolve, while the insoluble material can be removed while it floats, or is filtered from the solution. To extract the morphine from the processed opium, the latter is placed in a large volume of boiling water and calcium hydroxide added. The water is cooled and the unwanted alkaloids precipitate, while the morphine and some codeine remain in solution. The solution is then re-heated and ammonium chloride (and sometimes ethanol and diethyl ether) added. When the pH reaches 8−9, the

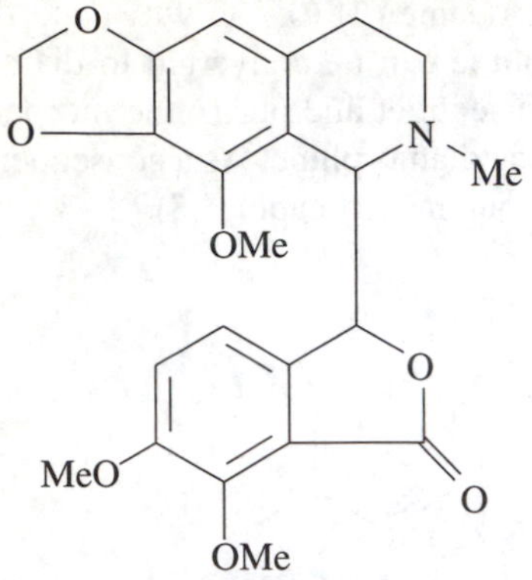

5 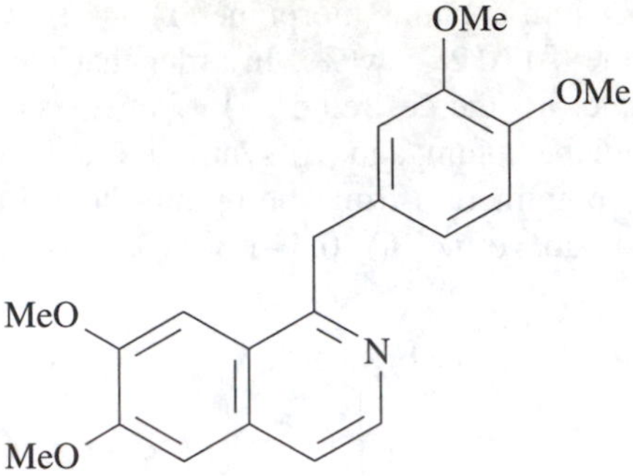   6

mixture is cooled and the morphine and any remaining codeine precipitates. The crude morphine is dried in the sun, ready for the next stage in the process.

DQ 5.1

Why do the opiate alkaloids precipitate from aqueous solution at pH 8–9?

Answer

At this pH, they are no longer present as salts and are therefore relatively apolar. Such materials are not freely soluble in water and as a consequence will precipitate.

Sometimes, the morphine is redissolved in dilute hydrochloric, sulfuric or tartaric acid, and activated charcoal added. The solution is heated and the charcoal filtered off. This process can be repeated a number of times. The filtrate is then dried, leaving the morphine in the salt form, as an off-white powder, or alternatively ammonium hydroxide may be added to the morphine solution, then precipitating the morphine from the aqueous solution as the free base.

DQ 5.2

What does this process achieve?

Answer

The addition of the acids causes the alkaloids to form salts. These salts are freely soluble in water. The addition of the charcoal can be used to decolourize the materials, by removal of the simple phenolics. Being insoluble, they are readily filtered from the mixture and the process can be repeated until the mixture is completely decolourized. Addition of the ammonium hydroxide basifies the solution and the opiates return to their free base forms, which are water-insoluble, hence forming precipitates which can be collected.

It is from the morphine that diamorphine is prepared. This is achieved by mixing the morphine with acetic anhydride and heating to approximately 85°C for about 5 h, or until all of the morphine has dissolved. Water is added to the mixture, followed by activated charcoal which absorbs any impurities. The mixture is repeatedly extracted with charcoal and filtered until the solution is clear. Sodium carbonate, dissolved in hot water, is slowly added to the mixture and the heroin base precipitates as a solid, which is then filtered and dried. The decolourizing and filtering process can be repeated a number of times until the desired colour/purity is achieved. From each kilogram of morphine, up to 700 g of diamorphine can be produced.

SAQ 5.1

Why is acetic anhydride used to acetylate the morphine?

To prepare the diamorphine hydrochloride, ethanol, diethyl ether and concentrated HCl are used. The base is dissolved in ethanol and the acid is added. Once all of the drug has been converted to the hydrochloride salt, further alcohol and ether are added. After a few minutes, crystals of heroin hydrochloride form. Further ether is added and the whole system allowed to stand. The solid diamorphine hydrochloride is then ready for packing and shipping. A consequence of all of these processes is that the opiates which may be contained within a heroin sample include diamorphine, 6-*O*-monoacetylmorphine (**7**), morphine, codeine and acetylcodeine (**8**).

7   8

5.3 Appearance of Heroin and Associated Paraphernalia

Heroin powder can exist in many different forms, from fine white powder, through to dark brown, granular lumps. The material may contain solvent from the manufacturing processes and larger samples typically smell of a mixture of acetic anhydride and acetic acid (like vinegar) due to the occluded solvent. The presence of the solvent may make the powder appear darker than its true colour, and this effect can be used as a 'tell-tale' sign that the sample might be heroin. In addition, the occluded solvents can be used to determine relationships between samples (summarized in Cole [3]).

Paraphernalia associated with the use of heroin may also be encountered. These items include metal spoons, materials used in the process of dissolution and heating heroin (candles, foil, etc.), materials used to filter the dissolved samples, such as cotton wool, and items used for the injection of the solubilized product. All such materials should be treated with care due to the problems associated with needle-borne diseases.

5.4 Bulk and Trace Sampling Procedures

With heroin samples, there are three problems that need to be addressed during the sampling process. These are (i) is the sample a bulk sample or a trace sample, (ii) how many of the items should be sampled, and (ii) how should the materials be described, sampled and analysed?

The issue of whether the material constitutes a bulk sample or a trace sample is relatively easily resolved. It may be considered that if the sample can easily be contaminated then it is a trace sample – otherwise it can be treated as a bulk sample. It should, however, be remembered that all possible precautions to minimize contamination of any sample should be taken at all times.

The second of the questions has given rise to a number of different studies. In general, the materials should be separated out into visually identical groups. The number of items in each group should then be determined. One model, proposed by the United Nations Drug Control Programme (UNDCP) [4] suggests that if the group size is less than 10, then each member of the group should be sampled, if the size is between 10 and 100 packages, then 10 packages from the group should be sampled, and if the size is greater than 100 items, then the square root of the number of items should be sampled. The items to be identified for sampling should be randomly chosen. This can be achieved by assigning each member of each group a number, starting at 1, and choosing the items to be sampled by using a random-number table or generator. There are other models available for determining how many items should be sampled, but a discussion of these is outside the scope of this present text. In addition to the theoretical models proposed, there may be requirements laid down by the legal jurisdiction in which the scientist is working and these should always be considered.

The issue of how a bulk sample should be described, sampled and analysed is equally problematic. The description itself is straightforward and should include all of the details relevant to a forensic science examination, including, for example, packaging, labelling, and the size, colour, weight, form and smell of the material. Sampling is a more difficult area. For small amounts of material,

Plate 4.1 Glandular trichromes found on the surface of *Cannabis sativa*. Copyright Michael D. Cole, Anglia Polytechnic University, Cambridge, UK, and reproduced with permission.

Plate 4.2 Unicellular trichromes found on the surface of *Cannabis sativa*. Copyright Michael D. Cole, Anglia Polytechnic University, Cambridge, UK, and reproduced with permission.

Plate 4.3 Cystolithic trichromes found on the surface of *Cannabis sativa*. Copyright Michael D. Cole, Anglia Polytechnic University, Cambridge, UK, and reproduced with permission.

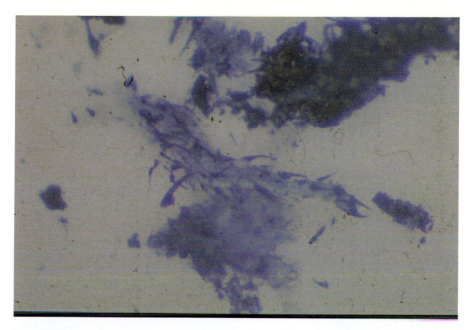

Plate 4.4 Unicellular trichromes present in the remains of smoked cannabis material. Copyright Michael D. Cole, Anglia Polytechnic University, Cambridge, UK, and reproduced with permission.

Plate 4.5 Wrapping materials, illustrating how a physical fit might be identified. Copyright Michael D. Cole, Anglia Polytechnic University, Cambridge, UK, and reproduced with permission.

Plate 4.6 A low-power microscopic image of cannabis resin, illuminated to show raised areas on one common surface and the corresponding troughs on the opposite surface. Copyright Michael D. Cole, Anglia Polytechnic University, Cambridge, UK, and reproduced with permission.

the cone-and-square method may be employed, but this is not practical for larger amounts of powdery materials. For such items, the use of a Waring blender has been advocated. For very large samples, it has been suggested that taking a core of the material is the best way to sample for analysis. The important factor is to ensure that the materials used for analysis are homogeneous.

SAQ 5.2

Why is it necessary to ensure that a material is homogeneous?

With items which might contain traces of drug, the principle concern is the avoidance of contamination. All of the laboratory materials, surfaces, chemicals and the worker must be demonstrably clean of drugs prior to commencing the analysis. The drugs should be recovered by using either a swab soaked in a suitable solvent and subsequently removing the drug from the swab, or by washing the item directly with solvent and then using the extract obtained. A suitable solvent is one which satisfies a number of criteria, including (i) not catalyzing the breakdown of the components of the heroin sample, (ii) not reacting with components of the heroin in subsequent analyses, and (iii) being compatible with the analytical systems and methods to be employed.

DQ 5.3

What would make a good solvent for the recovery of trace amounts of heroin?

Answer

Methanol is a good universal solvent, but may result in the rapid hydrolysis of trace amounts of diamorphine to monoacetylmorphine and morphine itself. As such, where possible, it should be avoided for heroin analysis, in particular the analysis of trace samples of heroin. Chloroform is an ideal solvent, but will rapidly evaporate from the swab. With care, however, it can be used since there are not the hydrolysis problems associated with methanol.

5.5 Identification, Quantification and Comparison of Heroin Samples

Whatever the type of sample, having carried out a description of the material and obtained a suitable sample for analysis, the next stage is the analytical process itself, i.e. to identify, quantify and, when necessary, make comparisons between samples. If the sample is a trace sample suspected of being heroin, the usual route is to proceed directly to GC–MS analysis. If the sample is a bulk sample, the sequence includes (i) presumptive tests, (ii) TLC (particularly when samples are to be compared), (iii) confirmation of identity, usually by GC–MS analysis (which can also be used for comparison purposes) and, if required, (iv) quantification and further sample comparison by HPLC.

5.5.1 Presumptive Tests for Heroin

Heroin samples, often identified in a preliminary way on the basis of their smell, comprise a large number of opiate drugs which are frequently adulterated with a wide variety of other drugs and pharmacologically inactive materials. Some of the drugs present in this way are themselves controlled substances and so a wide variety of presumptive tests need to be performed to identify all of the classes of drug that might be present.

The opiate drugs can be screened for by using a combination of tests (see Appendix 1), as shown in Table 5.1. What is also clear, however, is that each of the tests alone are not specific to one opiate drug. This means that further tests are required to determine which of the opiates are present.

In addition to the opiates themselves, other drugs may also be present. These include barbiturates (most commonly, phenobarbitone), procaine and lignocaine, caffeine, paracetamol (acetomenophen) and methaqualone. Some of these drugs are also controlled and should therefore be tested for. The barbiturates are tentatively identified by using the Dille–Koppanyi reagent, procaine and lignocaine with the cobalt isothiocyanate reagent, and methaqualone with a modified cobalt isothiocyanate test and the Fischer–Morris test (for further details of these reagents, see Appendix 1). If these drugs or drug classes are identified as being present, then further analysis will be required, as described elsewhere in this present book.

The situation is further complicated by the fact that different drugs may be mixed with heroin at different times. For example, in European heroin, in the early 1980s, the drugs were mainly adulterated with caffeine and procaine. Towards the mid to late 1980s, this changed to adulteration with phenobarbitone and

Table 5.1 Colour reactions observed on reacting opiates with common presumptive test reagents (see Appendix 1)

Colour test	Colour change observed
Marquis reagent	Opiates react to give blue to violet products on a colourless background
Mandelin reagent	Opiates react to give olive-green products on an orange background
Cobalt isothiocyanate	Diamorphine reacts to give a blue product on a pink background; the other opiates do not react. Also reacts with tropane alkaloids and some benzodiazepines

methaqualone, while in the 1990s, the use of these latter materials decreased and the addition of paracetamol and caffeine increased. The reasons for these changes are largely unexplained but may be related to availability of materials [5]. It is for this reason that the full suite of presumptive tests should be carried out on each sample to be analysed because it cannot be stated with any certainty what the diluents are without chemical analysis.

5.5.2 Thin Layer Chromatography of Heroin Samples

Having identified the materials as potentially containing opiates, the next phase in the analysis is to confirm the identity of which opiates are present and, where necessary, to carry out a comparison of samples. In order to identify which drugs are contained in a sample, without employing expensive instrumental techniques, thin layer chromatography can often be used.

SAQ 5.3

What information can TLC provide?

The samples for thin layer chromatography should be dissolved in a solvent which satisfies a number of criteria, and the solution then applied to the TLC plate.

DQ 5.4

What criteria should a good solvent for the application of a drug sample to a TLC plate meet?

Answer

A good solvent should (i) not react with the drugs being analysed (in this case, for example, with the hydroxide groups of some opiates), (ii) be volatile, to allow rapid application of a small spot of solvent and at the same time concentrate the sample prior to analysis, (iii) freely dissolve all of the drugs of interest, so that they are determined quantitatively and at the same time no solid material is present in the mixture immediately prior to analysis after removal of the insoluble adulterants, and (iv) be free of water, to prevent deactivation of the silica gel and to reduce the risk of sample hydrolysis during the process of sample application to the TLC plate.

Of the solvents available, chloroform has been found to satisfy these criteria for heroin analysis. The samples should be dissolved at a concentration which is of the same order of magnitude as the standards against which they are to be compared. TLC of heroin samples is most frequently carried out on silica gel. Between 5 and 10 μg of each component drug should be applied to the plate. The actual concentration of the solution prepared from a street drug sample will depend upon the drug concentration in the material, but typically 10 μl amounts of solutions prepared at $1-10$ mg ml^{-1} will allow visualization of the majority of compounds found in a typical heroin sample. The solution should be applied in as small a spot as possible and positive controls (opiates dissolved in chloroform at known concentrations, typically each at 1 mg ml^{-1}), and negative controls (chloroform alone), should be applied to the chromatographic plate. Having applied the mixtures to be separated, the plate is placed in a TLC developing tank, which has been pre-saturated with solvent, and then developed.

There are a number of different solvent systems which can be used to separate the components of heroin mixtures (Table 5.2). In order to identify which heroin components are present, it necessary to use at least two of these solvent systems. Standards (positive controls) and negative controls (blanks) should be analysed on each chromatographic plate, in addition to the samples themselves.

Having developed the chromatogram, it should be removed from the chromatographic tank, dried at room temperature and then visualized in (i) white light, (ii) short-wavelength ultraviolet (UV) light (254 nm), and (iii) long-wavelength UV light (360 nm), following spraying with a suitable developing reagent. These latter may include acidified potassium iodoplatinate and Dragendorff's reagent.[†] The data obtained should be recorded at each stage, including the colours and retardation factor (R_f) values of the compounds observed at each stage of the visualization process. By using this approach, it is possible to compare different samples, tentatively identify the opiate components of the heroin samples

Table 5.2 TLC solvent systems that can be used to separate opiate drugs and adulterants found in heroin samples and published R_f data obtained when using these systems

Drug	MeOH/NH$_3$ (880)[a]	Cyclohexane/toluene/Et$_2$NH[b]	CHCl$_3$/MeOH[c]
Components of heroin			
Diamorphine	0.47	0.15	0.38
6-*O*-Monoacetylmorphine	0.46	0.06	0.19
Morphine	0.37	0.00	0.09
Codeine	0.33	0.06	0.18
Adulterants			
Methaqualone	0.74	—	—
Caffeine	0.63	—	—
Procaine	0.69	—	—
Lignocaine	0.55	—	—

[a] 100:1.5, by volume.
[b] 75:15:10, by volume.
[c] 9:1, by volume.

[†] A solution of potassium iodide and bismuth subnitrate in acetic acid, used for the detection of alkaloids and quaternary nitrogen compounds.

(further methods described elsewhere in this book may be necessary to identify other drugs in different classes found in the specimens) and draw up a list of components required for the standard mix to be analysed to confirm the identity of the drug.

SAQ 5.4

Why can definitive identification of opiates not be achieved by TLC?

Having reached this stage, the next step is confirmation of the identities of the components of the mixture. This is usually achieved by using gas chromatography–mass spectrometry.

5.5.3 Gas Chromatographic–Mass Spectroscopic Identification of Heroin

The opiate drugs will (gas) chromatograph in both the derivatized and underivatized forms. In the underivatized drugs, free hydroxide moieties are present in 6-O-monoacetylmorphine, morphine and codeine. These form hydrogen-bonds with the analyte molecules and sorb strongly onto various components of the gas chromatographic system. These hydrogen-bonds are sufficiently strong that they cause problems such as tailing of the compounds and, as a consequence, derivatization of the drugs is advised. In addition to improving the chromatographic properties of the analytes, derivatization prevents the phenomenon of transacetylation, which has been observed when injections of underivatized drugs have been made into the chromatographic system when methanol is used as the solvent [6]. Derivatization can conveniently be achieved through the use of N,O-bis(trimethylsilyl)acetamide (N,O-BSA), although other reagents can be used. The following sequence can be employed for sample derivatization:

1. The sample is prepared in a suitable solvent, i.e. one which is volatile, free of water, does not react with the analyte(s) or catalyse their decomposition, and in which the analyte(s) is/are freely soluble.

2. The sample solution is centrifuged to remove any solid material which might block the syringe or components of the gas chromatographic system. The supernatant is then used for subsequent analysis.

3. The standards are prepared at a concentration of the same order of magnitude as the sample.

4. A solvent blank is also prepared for analysis.

   DQ 5.5

   Why is the blank necessary?

Answer

The blank is needed to demonstrate that the solvents and reagents do not contain compounds which co-elute with the internal standard and species of interest.

5. The samples and standards are placed in GC vials and 'blown down' under nitrogen. The latter gas is used to prevent sample decomposition.
6. The derivatization reagent (100 µl of 10% N,O-BSA in *n*-hexane containing 0.1 mg ml^{-1} of *n*-alkane internal standard) is added to the sample, the sample vial closed, the system shaken and then allowed to derivatize at room temperature.
7. The sample is then analysed, in the following order: derivatized blank, derivatized standard, derivatized blank, derivatized sample, derivatized blank..., etc.

SAQ 5.5
Why is this order of analysis employed?

N,O-BSA is a pre-column derivatizing reagent that reacts rapidly, at room temperature, with free hydroxide and amino moieties (Scheme 5.1 in the case of morphine), yielding products with good chromatographic properties.

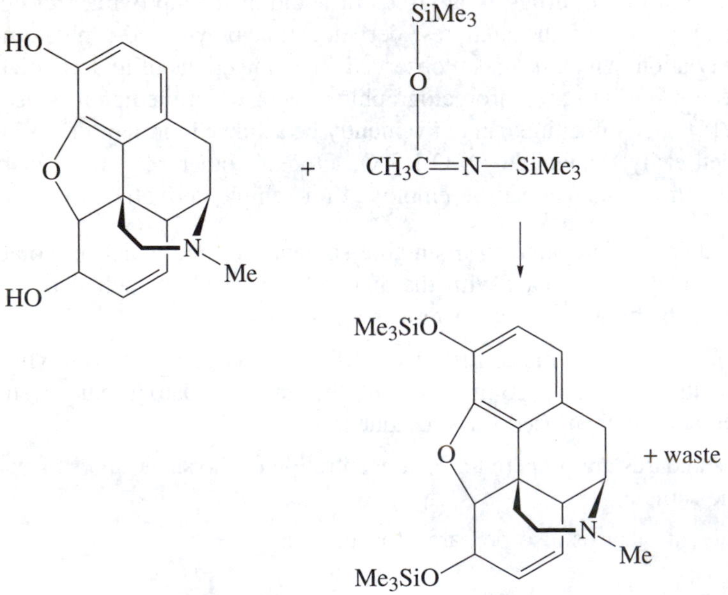

Scheme 5.1 Derivatization of morphine with *N,O*-bis(trimethylsilyl)acetamide.

Table 5.3 Typical GC operating conditions and parameters suitable for the analysis of opiate drugs by using GC-FID or GC–MS

System/parameter	Description/conditions
Column	OV-1: 25 m × 0.22 mm i.d.; d_f, 0.5 μm
Injection temperature	250°C
Column oven temperature programme	150°C for 2 min; increased to 280°C at 9°C min^{-1}; held for 2 min
Carrier gas	He, at a flow rate of 1 ml min^{-1}
Split ratio	50:1
Detector	Flame-ionization or mass spectrometric detection
Detector temperature (FID)	290°C
Derivatization reagent	N,O-bis(trimethylsilyl)acetamide

Typically, a temperature programme is used to separate the analytes. An example of the typical conditions used in such analyses is shown in Table 5.3, although it should be remembered that there will be variation between different laboratories and between the different instruments being used.

A typical chromatographic separation is presented in Figure 5.1. The compounds are identified in GC–MS analysis on the basis of two pieces of information, namely the retention time (or relative retention index) and mass spectral data.

Values of the *relative retention index*, calculated as follows:

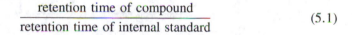

$$\frac{\text{retention time of compound}}{\text{retention time of internal standard}} \tag{5.1}$$

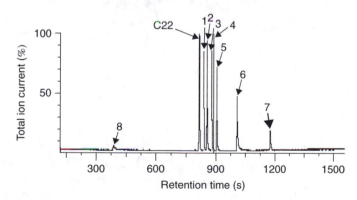

Figure 5.1 Separation of opiates derivatized with N,O-bis(trimethylsilyl)acetamide by GC–MS: 1, codeine; 2, acetylcodeine; 3, morphine; 4, 6-monoacetylmorphine; 5, diamorphine; 6, papaverine; 7, noscapine; 8, caffeine (C22 indicates the internal standard).

for each of the components of the drug sample are compared with those of the standards in a (standards) mixture. If a component in the drug and a specific standard have the same retention index, then a match is inferred. The mass spectra of the two compounds are subsequently compared. Since, in principle, under a given set of mass spectroscopic conditions, every compound will fragment in a unique and predictable way, if the mass spectra match then the identity of the compound is considered proven. Example of the mass spectra of diamorphine from a standard mixture and from a street sample of 'heroin' are shown in Figure 5.2.

However, it should be remembered that when chromatographing N,O-BSA derivatives, the mass spectra should be compared to those of standards which have been derivatized in the same way and *not* with the spectrum of the native drug.

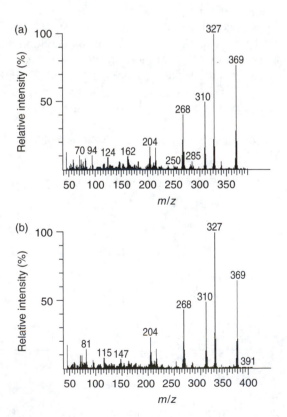

Figure 5.2 GC–MS analysis of diamorphine: electron-impact mass spectra of (a) a standard sample, and (b) that found in a street sample of 'heroin'.

SAQ 5.6

Why is this consideration necessary?

5.5.4 Quantification of Heroin Samples

The opiate drugs present in heroin can be quantified by using either GC or HPLC. When GC is used, the sample is often derivatized. Quantification when using this process makes the assumption that the sample has been completely and quantitatively derivatized. Such a treatment also precludes the problems associated with transacetylation [6] if the drugs are not treated in this way. In addition, however, the derivatization process adds further steps to the analysis that can result in sample breakdown and/or contamination. It is for these reasons that some laboratories carry out identifications of heroin by using GC–MS and then quantify the sample by employing HPLC. Examples of quantification using both GC and HPLC are discussed in the following sections.

5.5.4.1 Quantification of Heroin by GC

In this example, a gas chromatographic analysis was performed to determine the quantity of diamorphine in an illicit sample. The calibration data obtained are presented in Table 5.4.

The sample was dissolved at a concentration of 1 mg ml^{-1} in a suitable solvent. From the results obtained (see Table 5.5), it is possible to determine the quantity of diamorphine present in this sample, and provide the answer on a percentage basis. However, in order to establish that it is valid to use such data, a

Table 5.4 Calibration data, obtained by GC analysis, used for the determination of diamorphine in a 'heroin' sample

Diamorphine[a] concentration (mg ml^{-1})	Diamorphine peak area (arbitrary units)	Internal standard peak area (arbitrary units)
0.10	2399	21 047
0.10	2814	23 621
0.25	6545	22 529
0.25	6180	20 538
0.50	11 820	19 774
0.50	16 505	27 231
1.00	27 019	21 880
1.00	26 182	22 192
1.50	36 241	21 584
1.50	37 583	21 940
2.00	51 569	21 888
2.00	31 406	13 156

[a] As the free base.

Table 5.5 Data, obtained by GC analysis of a heroin sample, used for the determination of diamorphine

Replicate	Diamorphine[a] peak area (arbitrary units)	Internal standard peak area (arbitrary units)
1	8750	21 575
2	8821	22 010

[a] As the free base.

graph must first be plotted. The latter should be in the form of *relative response* (i.e. peak area of diamorphine/peak area of internal standard) against drug concentration. The data required for such a plot are given in Table 5.6, with the resulting graph shown in Figure 5.3.[†]

Using these data, it is possible to formulate a regression equation by employing the method of least-squares and solving the following simultaneous equations:

$$\sum Y = m \sum X + nc \tag{5.2}$$

$$\sum XY = m \sum X^2 + c \sum X \tag{5.3}$$

The summed values are shown in Table 5.6 and when these are substituted into the above equations and the latter solved, the regression equation, $Y = 1.168X + 0.007$,

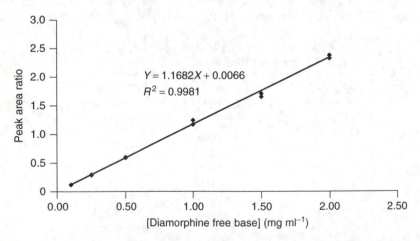

Figure 5.3 Calibration curve used for the determination of diamorphine in a 'heroin' sample by GC.

[†] R in this figure is known as the *correlation coefficient*, and provides a measure of the quality of calibration. In fact, R^2 is used because it is more sensitive to changes. This varies between -1 and $+1$, with values very close to -1 and $+1$ pointing to a very tight 'fit' of the calibration curve.

Table 5.6 Calibration and calculated data used for the determination of diamorphine in a 'heroin' sample

Diamorphine[a] concentration (mg ml^{-1})[b]	Diamorphine peak area (arbitrary units)	Internal standard peak area (arbitrary units)	Y-value[c]	X^2	$X \times Y$
0.10	2399	21 047	0.11	0.01	0.01
0.10	2814	23 621	0.12	0.01	0.01
0.25	6545	22 529	0.29	0.06	0.07
0.25	6180	20 538	0.30	0.06	0.08
0.50	11 820	19 774	0.60	0.25	0.30
0.50	16 505	27 231	0.61	0.25	0.30
1.00	27 019	21 880	1.23	1.00	1.23
1.00	26 182	22 192	1.18	1.00	1.18
1.50	36 241	21 584	1.68	2.25	2.52
1.50	37 583	21 940	1.71	2.25	2.57
2.00	51 569	21 888	2.36	4.00	4.71
2.00	31 406	13 156	2.39	4.00	4.77
10.70	**256 263**	**257 380**	**12.58**	**15.15**	**17.76**

[a] As the free base.
[b] X-value.
[c] Ratio of diamorphine peak area to internal standard peak area.

is obtained. In order to obtain the concentration of the drug, the ratios of the responses (GC peak areas) are calculated for the two replicates (see Table 5.5), giving values of 0.406 and 0.401, respectively. These are then averaged (0.4035) and this value is then substituted into the regression equation, giving a concentration of 0.34 mg ml^{-1}. Expressing this as a percentage of the original concentration (1 mg ml^{-1}), yields a final value of 34%. No correction for salt or free base is required in this evaluation.

5.5.4.2 Quantification of Heroin by HPLC
When quantifying opiates by HPLC, a number of basic principles should first be considered before the quantification process is undertaken.

DQ 5.6

What are good criteria for the solvent to be used for introduction of a heroin sample into an HPLC system?

Answer

First, if, for example, a powder is to be examined, then it should be freely soluble in the solvent chosen for injection into the chromatographic system. Secondly, the solvent should be fully miscible with the mobile phase. For these reasons, methanol is frequently chosen for heroin analysis, although the drug should not be left in this solvent for long periods because of the risk of hydrolysis of some of its components, e.g. monoacetylmorphine and diamorphine.

Baseline resolution of the compounds must be achieved in the chromatographic analysis so that a peak height or area can be assigned to one compound alone. In addition, it is important that the calibration curves in HPLC are produced from the same batch of solvent in which the sample is to be analysed. This is particularly important because small differences in pH can lead to different extinction coefficients when measuring UV absorptions, thus leading to inaccuracies in the quantification process.

When preparing a calibration curve, a wide enough range of concentrations should be chosen in order to ensure that the *sample* concentration will fall in the linear range of such a curve. This is especially true for heroin where a wide range of drug concentrations might be encountered in street samples.

Furthermore, when preparing the calibration curve, if the two-point or regression methods are employed, the solutions should be injected starting with the lowest concentration, then increasing to the highest. This reduces the risk of column priming. Between each sample solution, an injection of the solvent alone should be made. This ensures that the chromatographic system is clear of any contamination which may give rise to inaccurate results.

The following set of conditions have been shown to be efficient in the HPLC quantification of heroin:

Column:	Silica gel, 12.5 cm × 4.6 mm i.d.
Eluant:	isooctane/diethyl ether/methanol/water/diethylamine (40:325:225:15:0.65, by volume)
Flow rate:	2 ml min^{-1}
Detection:	UV at 230 nm

A typical HPLC separation of opiates which can be achieved under these conditions is shown in Figure 5.4.

In addition to retention time, if diode-array detection is employed, further confirmation of each of the eluted compounds can be achieved by observing the ultraviolet spectra obtained for the samples and standards.

SAQ 5.7

How can diode-array detection assist in analyte identification?

Single-point, two-point and linear regression methods can all be used for quantification [7]. Of these, the linear regression method is the most reliable, and is illustrated here.

A drug sample, found to contain diamorphine, has been quantified by HPLC. The calibration data obtained are shown in Table 5.7.

The sample, known to contain diamorphine, gave peak areas of 115 604 and 115 998 (in arbitrary units) for two replicated samples. In this case, we wish to calculate the percentage of diamorphine in the (street) sample which dissolved at a concentration of 2.5 mg ml^{-1}.

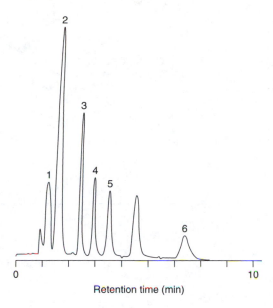

Figure 5.4 HPLC separation of opiates: 1, noscapine; 2, papaverine; 3, acetylcodeine; 4, diamorphine; 5, 6-monoacetylmorphine; 6, codeine; 7, morphine.

Table 5.7 Calibration data, obtained by HPLC analysis, used for the quantification of a 'heroin' sample

Drug concentration ($mg\,ml^{-1}$)	Peak height (arbitrary units)
0.10	25 828
0.10	15 363
0.25	51 518
0.25	60 544
0.50	91 661
0.50	91 897
1.00	175 403
1.00	175 421
1.50	248 601
1.50	264 063
2.00	341 032
2.00	340 956

This problem is solved as follows. In order to confirm the validity of the calibration data for quantification, a graph of response (i.e. peak height) against drug concentration is plotted (Figure 5.5). If the data lie on a straight line, then they can be used for quantification.

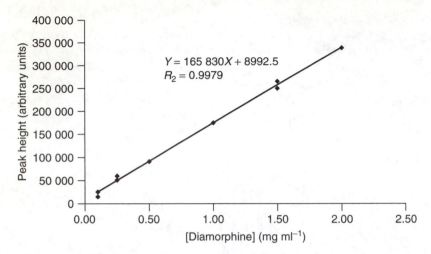

Figure 5.5 Calibration curve used for the determination of diamorphine in a 'heroin' sample by HPLC.

The required regression equation is obtained by solving the following simultaneous equations:

$$\sum Y = m \sum X + nc \qquad (5.4)$$

$$\sum XY = m \sum X^2 + c \sum X \qquad (5.5)$$

using the data given in Table 5.8.

When these values are substituted into the above equations and the latter solved, we obtain the regression equation, $Y = 165\,830X + 8992$. In order to calculate the amount of diamorphine in the sample, the data values are averaged (115 801) and then substituted into the regression equation. This yields a concentration of 0.64 mg ml^{-1}. However, the latter needs to be expressed as a percentage of the original concentration (2.5 mg ml^{-1}) – this gives a final value of 25.8% diamorphine in the sample.

In some cases, the calibration data are provided with the diamorphine in salt form (usually as the hydrochloride). Under such circumstances, it is better to calculate the amount of base present in the sample, since it will not be known which particular salt form the diamorphine will exist as for the sample being quantified.

5.5.5 Comparison of Heroin Samples

It is sometimes necessary for a drugs chemist to make comparisons of heroin samples. While TLC, HPLC and GC (i.e. GC–MS) can all be employed for this purpose, there are advantages and disadvantages to each of these techniques.

Table 5.8 Calibration and calculated data used for the determination of diamorphine in a 'heroin' sample

Drug concentration (mg ml^{-1})^a	Peak height (arbitrary units)^b	X^2	$X \times Y$
0.10	25 828	0.01	2582.8
0.10	15 363	0.01	1536.3
0.25	51 518	0.06	12 879.5
0.25	60 544	0.06	15 136
0.50	91 661	0.25	45 830.5
0.50	91 897	0.25	45 948.5
1.00	175 403	1.00	175 403
1.00	175 421	1.00	175 421
1.50	248 601	2.25	372 901.5
1.50	264 063	2.25	396 094.5
2.00	341 032	4.00	682 064
2.00	340 956	4.00	681 912
10.70	**1 882 287.00**	**15.15**	**2 607 709.60**

^{a} X-value.
^{b} Y-value.

DQ 5.7

What are the relative advantages and disadvantages of TLC, HPLC and GC (or GC–MS) for heroin comparisons?

Answer

*TLC is not quantitative and while it is a rapid and inexpensive means of obtaining **exclusion** of any relationship between heroin samples, it cannot be used with any reliability to confirm that two samples were once part of the same larger batch. HPLC is quantitative but does suffer from lack of definitive identification of each (chromatographic) peak. It is for these reasons that many forensic science laboratories use GC–MS for the comparison of heroin samples.*

In such an investigation, the samples are treated and analysed in exactly the same way as that used for identification of the materials. If the resulting chromatograms (Figure 5.6) and the mass spectra cannot be differentiated, then 'a match is called' between the samples, as shown from this figure.

The data obtained can be used in a number of different ways. At a simple level, two samples of heroin may be compared, as shown in Figure 5.6 for 'Heroin Bags 1 and 2'. Alternatively, the 'profile' of the heroin being examined may indicated geographic origins and relationships (or otherwise) to other batches which have been seized. A full discussion of such comparisons is beyond the scope of this present book, but the interested reader is directed towards recent studies in this area [6, 8, 9].

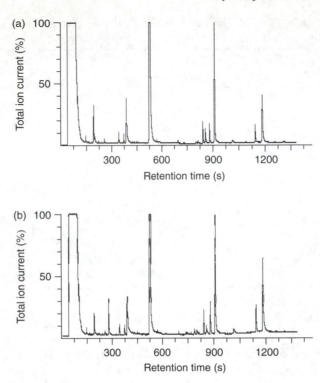

Figure 5.6 Total ion chromatograms of two different 'heroin' samples: (a) 'Heroin Bag 1'; (b) 'Heroin Bag 2'. The identical forms of the two traces allows the analyst to 'call a match' between the two samples.

Summary

Heroin can be encountered in the form of powder samples and as residues on the paraphernalia used for its administration and on the equipment involved in its mixing, cutting and distribution. 'Heroin' is the mixture of products formed when morphine is extracted from the dried latex of the field poppy, purified (often incompletely) and then subsequently acetylated to form diamorphine, the target drug.

Bulk samples are analysed by physical description, presumptive tests and thin layer chromatography, followed by confirmation using GC–MS. HPLC is generally used for sample quantification. Trace samples are analysed by the direct examination of recovered materials employing GC–MS.

Sample comparison can be achieved by using TLC, followed by HPLC employing diode-array detection, or by GC–MS. TLC, while lacking sensitivity and also being a qualitative method, can be used, however, to achieve discrimination between samples where these are clearly different. If such distinction between samples cannot be achieved by this technique, then GC–MS or HPLC (with diode-array detection) should be used. GC–MS, however, offers greater resolution and powers of sample identification than HPLC.

References

1. Fulton, C., *The Opium Poppy and other Poppies*, US Treasury Department, Bureau of Narcotics, US Government Printing Office, Washington, DC, 1944, pp. 37–38.
2. Farnilo, C. G., Rhodes, H. L. J., Hart, H. R. L. and Taylor, H., 'Detection of morphine in *Papaver setigerum* DC', *Bull. Narcotics*, **5**(1), 26–31 (1953).
3. Cole, M. D., 'Occluded solvent analysis as a basis for heroin and cocaine sample differentiation', *Forensic Sci. Rev.*, **10**, 113–120 (1998).
4. United Nations Drug Control Programme, *Recommended Methods for Testing Heroin*, Manual for use by national narcotics laboratories, United Nations Division of Narcotic Drugs, New York, 1988.
5. Kaa, E., 'Impurities, adulterants and diluents of illicit heroin. Changes during a 12 year period', *Forensic Sci. Int.*, **64**, 171–179 (1994).
6. Gough, T. A., 'The examination of drugs in smuggling offences', in *The Analysis of Drugs of Abuse*, Gough, T. A. (Ed.), Wiley, Chichester, UK, 1991, pp. 515–529.
7. Cole, M. D. and Caddy, B., *The Analysis of Drugs of Abuse: An Instruction Manual*, Ellis Horwood, Chichester, UK, 1995, pp. 94–100.
8. Janhunen, K. and Cole, M. D., 'Development of a predictive model for batch membership of street samples of heroin', *Forensic Sci. Int.*, **102**, 1–11 (1999).
9. Strömberg, L., Lundberg, L., Neumann, H., Bobon, B., Huizer, H. and van der Stelt, N. W., 'Heroin impurity profiling: A harmonisation study for retrospective comparisons', *Forensic Sci. Int.*, **114**, 67–88 (2000).

Chapter 6
Cocaine

Learning Objectives

- To be aware of the origins and the manufacturing processes used to synthesize cocaine.
- To understand how cocaine can be identified.
- To learn how cocaine can be quantified.
- To appreciate that cocaine samples can be profiled.
- To provide the correct interpretation following the analysis of cocaine samples.
- To appreciate the difficulties associated with these methods.

6.1 Introduction

There is anecdotal evidence that cultivation of 'Coca leaf' from *Erythroxylum spp.* has occurred for the last 20 000 years, although the first archaeological evidence appears to occur at around 3000 BC [1]. The species *Erythroxylum spp.* are of significance because they provide the raw material for the production of cocaine (**1**). Of the approximately 200 species of *Erythroxylon*, 17 members of the genus are reported to produce cocaine, but of greatest significance are *E. coca* Lam. and *E. novogranatense* (Morris) Hieron., the cultivated varieties of which produce the highest levels of this drug. On average, these levels lie between 0.5 and 1.5% of the dry weight of the leaf material [2]. In truly wild varieties, the levels of cocaine are much lower at around 0.0005 wt% of the leaf material [3]. The dried plant materials are either chewed or ingested directly, or a tea can be prepared from them.

Cocaine is produced from alkaloids obtained from the plant *Erythroxylon coca*. Since the drug is produced in a batch process, the samples are variable, with no two of them being identical. Coca leaf itself can be chewed as a drug, or 'coca paste' or cocaine can be produced. Coca paste is an off-white or beige coloured powder comprising damp friable aggregates. Cocaine, as the hydrochloride salt (sometimes known as 'snow'), is itself a white, or off-white, powder, which typically smells of HCl. This salt may be re-converted to the free base form – which is more volatile – for smoking, and in doing so, hard, waxy lumps are produced which are known as 'rocks' or 'crack'.

In the United Kingdom, cocaine is listed as a Class A drug in Schedule 2 to the Misuse of Drugs Act, 1971, as are ecgonine (2), benzoyl ecgonine (3), ecgonine methyl ester (4) and any derivative of ecgonine which is convertible to ecgonine or cocaine. Under United Kingdom legislation, the isomer(s) involved do not have to be defined. In addition, coca leaf is also a Class A drug. However, its control was modified by the Misuse of Drugs Regulations, 1985. Although material containing 0.1% or less of cocaine is still, by definition, a Class A drug, through Schedule 5 of the regulations it is lawful for anyone to possess such material. This is because the concentration of the cocaine in this case is so low that it is deemed to be harmless [4]. In the United States, cocaine and crack cocaine are both classified as Schedule II drugs.

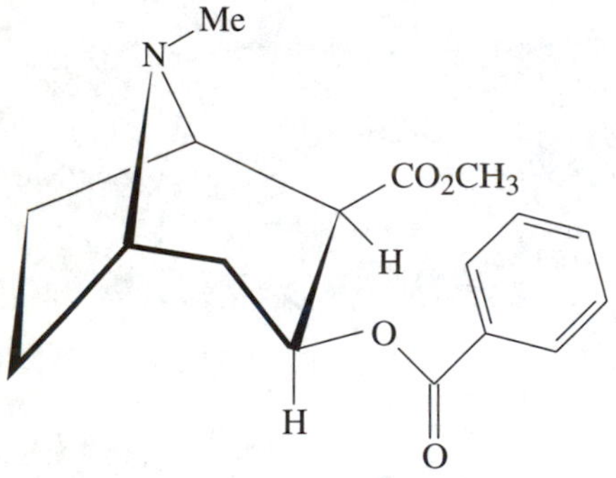

1

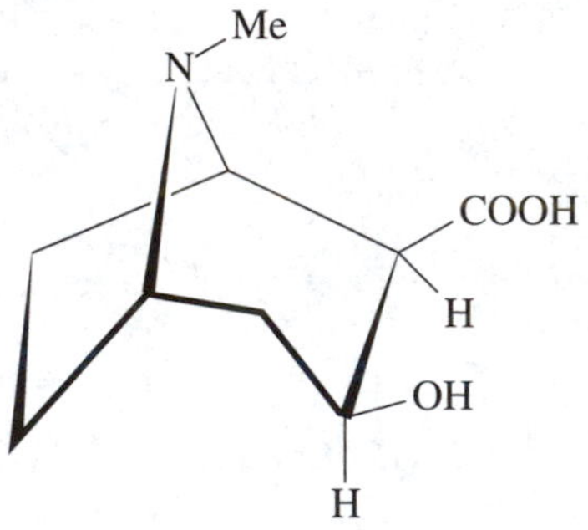

2

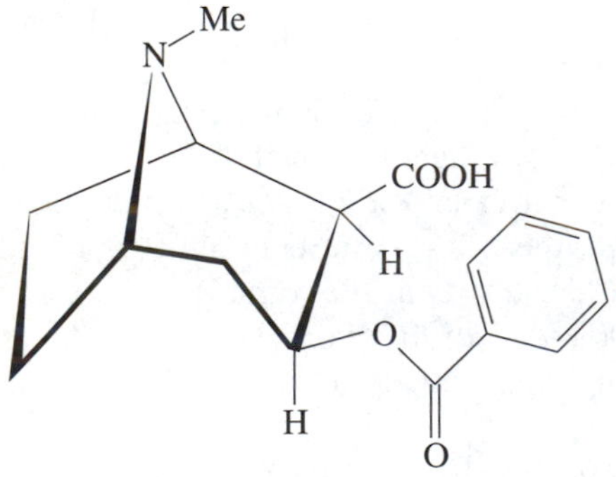

3

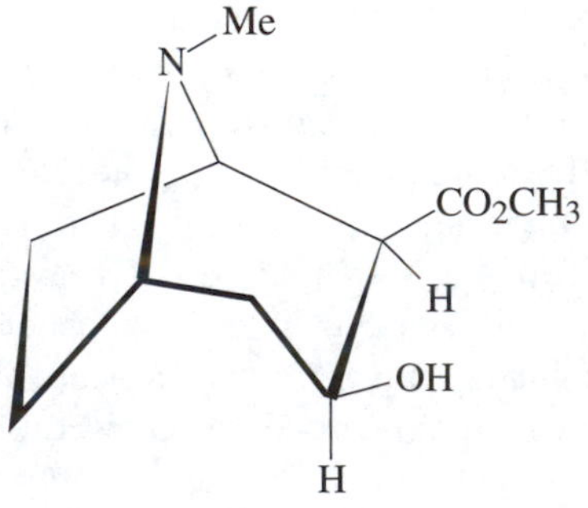

4

Cocaine can be administered via a variety of different routes. Popularly, the hydrochloride salt is insufflacated ('snorted') from a line of white powder and the drug absorbed across the mucous membranes of the nose. Alternatively, it may be administered by injection. 'Crack' cocaine, or 'Rocks' may be administered by smoking.

The amount taken depends upon the degree of dependency on the drug and the frequency of use of the drug by the user, as does the effect. Those effects which are sought include increased alertness, wakefulness, mood elevation, euphoria, an increase in athletic performance, a decrease in fatigue, clearer thinking, concentration, and increases in energy. Unwanted side-effects include increased irritability, insomnia, and restlessness. With high doses, a user can exhibit confused and disorganized behaviour, irritability, fear, paranoia, hallucinations, and may become extremely antisocial and aggressive, possibly leading to stroke, heart attack or death.

6.2 Origins, Sources and Manufacture of Cocaine

Although cocaine occurs naturally in plant material, a number of related, naturally occurring alkaloids based upon the tropane structure may be isolated from the plant material and subsequently converted into cocaine. These include ecgonine, ecgonine methyl ester and benzoyl ecgonine There are two principle routes by which extraction can be achieved prior to cocaine synthesis.

6.2.1 Extraction and Preparation of Coca Paste

Following cultivation of the plant material, the leaves from which the cocaine will be prepared are harvested and dried in the sun. From these, coca paste and, subsequently, cocaine is produced. In general, coca paste is prepared by one of two methods. The first involves wetting the leaves and macerating them with dilute sulfuric acid, thus forming the water-soluble sulfate salts of the alkaloids. The mixture is then extracted with kerosene. After phase separation, the aqueous layer is basified with ammonia, lime or sodium carbonate, and the alkaloids precipitated. They are then recovered by filtration.

SAQ 6.1

Why is a base used to precipitate the alkaloids?

The second method involves basifying the leaves with sodium or potassium carbonate, macerating the leaves, and adding kerosene, into which the alkaloids are extracted. Dilute aqueous sulfuric acid is then used to collect the alkaloids as the sulfate salts. The aqueous layer is then basified and the alkaloids filtered and recovered.

DQ 6.1

What is the chemical basis underlying this second method of alkaloid collection?

Answer

The basification of the mixture results in the alkaloids being present in their free base forms and hence they are soluble in apolar organic solvents (in this case, kerosene) into which they extract. Addition of the aqueous acid results in the sulfate form of the drug being produced, which preferentially partitions into the aqueous phase. Addition of the base neutralizes the acid, returning the drugs to their apolar, free base forms, which are not soluble in water, and hence will precipitate.

6.2.2 Synthesis of Pure Cocaine

The tropane alkaloids extracted from the leaves can be used to prepare pure cocaine. One route which is possible is as follows. The alkaloids are hydrolysed by using 1 M hydrochloric acid, which produces ecgonine. The latter is treated with 10% boron trichloride in methanol, to give ecgonine methyl ester, which is then reacted with benzoyl chloride, producing cocaine.

6.3 Qualitative Identification of Cocaine

In terms of gross morphology, characteristically, coca leaf has two lines which run parallel to the mid-rib on the underside of the leaf. Taxonomically, however, it is difficult to identify leaf and plant material to species on the basis of morphology (leaf and flower structure) alone and indeed hybridization between species is common [5]. In order to establish that the material comes from the genus *Erythroxylon* and contains controlled substances, it is therefore necessary to demonstrate the presence of cocaine. In bulk (those which can be seen by the naked eye) samples, this is achieved through a good physical description of the materials and packaging, followed by a combination of presumptive tests, TLC and a tandem technique, usually GC–MS.

6.3.1 Presumptive Tests for Cocaine

There are a number of presumptive tests for cocaine available from the literature. These are described below.

6.3.1.1 The Cobalt Isothiocyanate Test

Although there are several presumptive tests available for cocaine and related compounds, none alone is specific to cocaine itself. A very simple test for cocaine is the addition of the material under investigation to a 2% (wt/vol) solution of cobalt isothiocyanate in water. The presence of cocaine and related alkaloids results in a blue colour being formed. Interestingly, diamorphine and temazepam also result in a blue reaction, but none of the other opiates or benzodiazepines result in this colour reaction. In addition, a number of local anaesthetics also produce a positive colour reaction. It should be remembered that positive and negative controls should always be undertaken at the same time as the sample is being examined.

DQ 6.2

What is the purpose of the positive and negative controls?

Answer

*The negative control is undertaken to demonstrate that the equipment is clean and that any colour change observed is due to the **actual** presence of the drug. The positive control demonstrates that the test is working and provides a reference colour against which the reaction due to the sample may be compared.*

6.3.1.2 The Scott Test

This is a modification of the cobalt isothiocyanate test, and involves a 2% (wt/vol) solution of cobalt isothiocyanate in water, diluted with an equal volume of glycerine (Reagent 1), concentrated hydrochloric acid (Reagent 2) and chloroform (Reagent 3). In order to test the sample for the presence of cocaine, a small amount of material is placed in a test-tube and 5 drops of the first reagent are added. A blue colour develops if cocaine is present. One drop of concentrated hydrochloric acid is then added and the blue colour, if it results from the presence of cocaine, should disappear, leaving a pink solution. Several drops of chloroform should then be added. An intense blue colour forms in the chloroform (lower) layer if cocaine is present. Again, positive and negative controls should be performed concurrently with analysis of the test sample.

DQ 6.3

What advantage does this modified test have over the cobalt isothiocyanate approach?

Answer

This test allows greater discrimination. It is known that a number of products form blue complexes with the cobalt isothiocyanate, although these do not extract into the chloroform as readily, or if at all.

6.3.1.3 The Methyl Benzoate Odour Test

This test relies on the hydrolysis of the benzoate ester of cocaine and benzoyl ecgonine by potassium or sodium hydroxide [6]. A 5% (wt/vol) solution of the hydroxide is prepared in methanol. The sample under test is mixed with a small volume of this reagent in a suitable container and warmed gently, for example, on the palm of the hand. The excess alcohol is allowed to evaporate and the smell arising from the mixture noted (very great care should be exercised when assessing this). A positive and a negative control should also be undertaken. The resulting smell is that of methyl benzoate (familiar in the smell of oil of wintergreen). Piperocaine and benzoylecgonine (both benzoate esters) also result in a positive test.

SAQ 6.2

What precautions should be taken with this test?

If positive tests are obtained in the presumptive tests, the next stage in the analysis is to undertake TLC or to proceed directly to a spectroscopic evaluation. The use of TLC will allow a rapid, inexpensive means of discriminating samples containing cocaine from the false positives and a determination of what should be used in the standard mixtures for GC–MS analyses.

6.3.2 Thin Layer Chromatography

Prior to the use of any further method for the analysis of the material and identification of cocaine, the latter must be extracted. For powdered samples, which have been thoroughly homogenized, a 1 mg ml^{-1} solution in methanol will suffice. For herbal material, the situation is rather more complex if the problems associated with extraction of chlorophyll, etc. are to be avoided. One study, where methanol and ethanol alone and in the presence of either diethylamine or ammonia were used, demonstrated that refluxing in ethanol for 15 min resulted in a 93% recovery of the cocaine from the leaf material. This investigation further demonstrated that using an acid back-extraction, followed by a chloroform extraction, could allow the extract to be 'cleaned up' sufficiently enough for good, quantitative chromatography by packed-column GC to be achieved [2]. However, care should be used when undertaking a cocaine extraction. In aqueous solutions, if the pH is above 5.5, decomposition of the cocaine to benzoyl ecgonine and methanol will occur. At pH 5.8, a 10% loss is observed after 13 d, while at pH 8, 35% of the cocaine is lost after 4.5 h [5].

The presence or absence of cocaine in a sample may qualitatively be determined by using TLC, carried out on silica gel. A number of different solvent systems are available for this (Table 6.1). The compounds can be examined visually

Table 6.1 TLC R_f data for cocaine and local anaesthetics obtained under the operating conditions described in the text

Compound	Solvent system	
	MeOH/NH$_3$ (100:1.5)[a]	Cyclohexane/toluene/Et$_2$NH (75:15:10)[a]
Cocaine	0.59	0.56
Ecgonine	0.84[b]	0.00
Ecgonine methyl ester	0.65[b]	0.44
Benzoyl ecgonine	0.25[b]	0.00
Lidocaine	0.69	0.40–0.55[b]
Procaine	0.55	0.08–0.16[b]

[a] By volume.
[b] Streaking observed, i.e. not a rounded spot.

under (i) white light, (ii) short-wavelength UV light (254 nm), and (iii) long-wavelength UV light (360 nm), and after using visualization reagents. These include general reagents such as 1% KMnO$_4$ in water, those used for other drugs of abuse, for example, acidified potassium iodoplatinate, or those used for alkaloid detection, for example, Dragendorff's reagent (see Section 5.5.2 above). As before, it should be remembered that positive and negative controls should be included in every chromatographic investigation.

DQ 6.4

What are the difficulties associated with interpreting data from TLC?

Answer

There are two principle problems. The first is lack of resolution, for example, in the methanol/ammonia system, cocaine/procaine and ecgonine methyl ester/lidocaine are not well resolved and yet may occur together in the mixture. A further difficulty is added by the lack of specificity of the visualization reagents. None of the reagents used here are specific to this group of compounds.

By comparison of the R_f data and the colours observed under different light conditions and with different visualization reagents, the compounds thought to be present in the mixture can be determined and standard mixes prepared for GC–MS analysis. In addition, it may be possible to make comparisons between samples.

6.3.3 Definitive Identification of Cocaine

GC–MS is routinely applied to the identification of cocaine. This is because the technique is both sensitive and of very high resolution. However, this method can also be used on simple mixtures. Trace samples should be obtained by swabbing the surfaces to be examined and subsequently processing the swabs by eluting the drug using methanol or ethanol. Where bulk samples are to be analysed, the sample should be thoroughly homogenized in preparation for this. Derivatization is frequently used and a number of reagents are available. The materials are processed as follows:

1. The sample is prepared in a suitable solvent, i.e. one which is volatile, free of water and does not react with the analytes, or catalyse their decomposition, and in which the analytes are freely soluble. Methanol or ethanol are generally used for cocaine samples.

2. The sample system is centrifuged to remove any solid material which might block the syringe or the gas chromatographic system. The supernatant is used for subsequent analysis.

3. The standards are prepared at a concentration of the same order of magnitude as the sample.

4. A solvent blank is also prepared for analysis.

5. The sample and standards are placed in GC vials and 'blown down' under nitrogen. The latter is used to prevent sample decomposition.

6. The derivatization reagent (a typical mixture comprises 100 μl of 10% N,O-BSA in *n*-hexane containing 0.1 mg ml^{-1} *n*-alkane internal standard) is added to the sample, the sample vial closed and shaken, and the system allowed to derivatize at room temperature.

7. The sample is then analysed in the following order: derivatized blank, derivatized standard, derivatized blank, derivatized sample, derivatized blank, ..., etc.

DQ 6.5

Why is derivatization recommended?

Answer

Such treatment is suggested because of the chemistry of the drugs in question. The ecgonines (cocaine does not derivatize) have free hydroxide and/or carboxylic acid moieties. These will sorb onto active and 'dirty' sites within the GC–MS system and also hydrogen-bond with each other in the gas phase in the chromatographic system. This will result in poor mass transfer and hence poor chromatographic behaviour, as observed through tailing peaks. By derivatizing these compounds, the hydroxide and carboxylic acid groups are protected and the problem is thus reduced.

Table 6.2 Typical GC operating conditions and parameters suitable for the analysis of cocaine by using GC-FID or GC–MS

System/parameter	Description/conditions
Column	OV-1: 25 m × 0.222 mm i.d.; d_f, 0.5 μm
Injection temperature	300°C
Column oven temperature programme	170°C for 2 mins; increased to 280°C at 16°C min^{-1}; held for 2 min
Carrier gas	He[a] or N_2[b], at a flow rate of 1 ml min^{-1}
Split ratio	50:1
Detector	Flame-ionization or mass spectrometric detection
Detector temperature (FID)	300°C
Derivatization reagent	N,O-(bistrimethylsilyl)acetamide

[a]With mass spectrometric detection.
[b]With flame-ionization detection.

Table 6.2 presents a typical set of gas chromatographic operating conditions which have been found to give efficient separation behaviour for samples of cocaine.

The retention times (or relative retention indices) and mass spectra of each component of the standard mix and of the samples are then compared. If both of these data types match, the identity of the compound can be inferred.

DQ 6.6

Figure 6.1 shows a chromatographic separation of eicosane and cocaine. What is the relative retention index of cocaine?

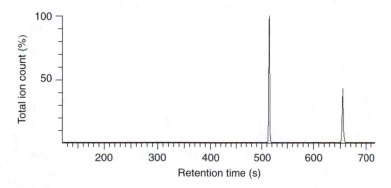

Figure 6.1 Chromatographic separation of eicosane, $C_{20}H_{42}$ (internal standard; t_R, 8.36 min) and cocaine (t_R, 10.56 min) (DQ 6.6).

Answer

This index is given by the ratio of the two retention times, in this case:

$$t_R(cocaine)/t_R(eicosane) = 10.56/8.36 = 1.26$$

DQ 6.7

Does it matter that the internal standard elutes before the analyte of interest?

Answer

No, this is not a problem, provided that the internal standard complies with a number of criteria. These include that it should not co-elute, react or catalyse the breakdown of other compounds that may be present, and that it exhibits good chromatographic behaviour in the system being used.

DQ 6.8

Figure 6.2 shows the mass spectrum of cocaine, following GC–MS analysis of this drug using an ion-trap detector. Explain each of the ions observed at m/z 304, 303, 272, 198, 182 and 82.

Answer

The molecular weight of cocaine is 303. In ion-trap detectors, protonated parent ions often form and so the ion at m/z 304 is $[M + H]^+$ (the ion at m/z 303 is the parent ion $[M]^+$). The ion at m/z 272 arises from $[M − MeO]^+$, while similarly the ion at m/z 198 represents $[M − C_6H_5CO]^+$. In addition,

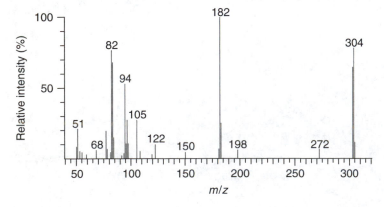

Figure 6.2 Electron-impact mass spectrum of cocaine, obtained by employing an ion-trap detector (DQ 6.8).

the ion at m/z *182 emanates from* $[M - C_6H_5COO]^+$, *and that at* m/z *82 results from the following fragment:*

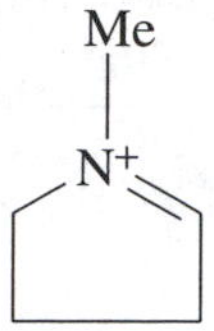

6.4 Quantification of Cocaine

While there are several methods available for the HPLC analysis of cocaine [7], these appear to have been principally used in toxicological studies. For street drug analysis, the preferred method is currently GC (either GC-FID or GC–MS). In addition to identification, GC–MS can also be used to quantify cocaine in street samples. For simple mixtures which contain only cocaine and a sugar, UV spectroscopy can also be employed for quantification purposes. Examples of each of these approaches are detailed below.

6.4.1 Quantification of Cocaine by GC–MS

This method relies on the response of the detector being in direct proportion to the concentration of drug in the sample being evaluated. A series of standards of known concentrations is prepared and analysed, with each analysis being separated by a blank.

SAQ 6.3

Why should each sample be separated by a blank in the analytical process?

The same solution system must be used to prepare both the standards and the samples.

SAQ 6.4

Why are internal standards necessary?

SAQ 6.5

Why is there a requirement to use the same solution for the preparation of the standards and the samples?

Table 6.3 Calibration and calculated data, obtained by GC–MS analysis, used for the quantification of cocaine in a drug sample

Cocaine[a] concentration (mg ml^{-1})[b]	Cocaine peak area (arbitrary units)	Internal standard peak area (arbitrary units)	Y-value[c]	X^2	$X \times Y$
0.188	280 959	8 062 322	0.035	0.035	0.007
0.188	309 454	9 364 403	0.033	0.035	0.006
0.375	779 646	9 165 776	0.085	0.141	0.032
0.375	764 902	9 423 551	0.081	0.141	0.030
0.750	1 184 352	9 411 496	0.126	0.563	0.094
0.750	1 390 950	10 782 244	0.129	0.563	0.097
1.500	2 007 799	9 231 358	0.217	2.250	0.326
1.500	2 107 793	9 542 613	0.221	2.250	0.331
3.000	3 002 575	6 672 401	0.450	9.000	1.350
3.000	4 129 815	9 572 136	0.431	9.000	1.294
11.625	**8 825 855**	**91 228 300**	**1.809**	**23.977**	**3.568**

[a] As the free base.
[b] X-value.
[c] Ratio of cocaine peak area to internal standard peak area.

Next, the peak areas of the drugs in the standards are obtained, and the relative responses calculated. Some exemplar standard data for the quantification of cocaine in a drug sample are presented in Table 6.3. Using these, a calibration graph of relative response against cocaine concentration, as the free base, is plotted (Figure 6.3), and from the simultaneous equations:

$$\sum Y = m \sum X + nc \tag{6.1}$$

$$\sum XY = m \sum X^2 + c \sum X \tag{6.2}$$

the appropriate regression equation can be calculated.

For the example being considered here, this equation is as follows:

$$Y = 0.1401X + 0.0181 \tag{6.3}$$

and therefore, using the HPLC data obtained from the drug sample, the concentration of cocaine can be calculated.

6.4.2 Quantification of Cocaine by UV Spectroscopy

This technique works particularly well if the cocaine is mixed with sugars, from which it can be easily separated by dissolution. Calibration standards are prepared, in methanol solution, at concentrations of 1.0, 0.5, 0.25, 0.125, 0.0625, 0.031 25 and 0.0156 mg ml^{-1}, using serial dilutions. A UV spectrum of a holmium filter is first measured to confirm that the instrumental set-up is performing satisfactorily.

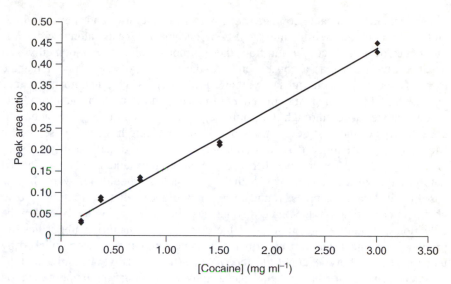

Figure 6.3 Calibration curve used for the determination of cocaine in a sample by GC–MS.

A baseline spectrum for the cuvette in which the cocaine is to be measured is then obtained by using methanol alone. Following this, at each cocaine concentration, ultraviolet spectra between 210 nm and 350 nm, measuring the absorbance values at λ_{max}, are recorded. Between each measurement, the cuvette is washed clean with 'HPLC-grade' methanol and dried. Using the absorbance values measured for the calibration standards, a curve of absorbance as a function of concentration of the cocaine free base is produced and the corresponding regression equation calculated. Once linearity of the data set has been confirmed, a solution of the material in which the amount of cocaine is to be determined, again at a concentration of 1 mg ml^{-1}, in the same methanol as the standards, is prepared (an ultrasonic bath is often used for this purpose). Any remaining solid material is then removed by centrifugation and the supernatant kept for further measurement (a methanol blank should also be prepared in the same way). An ultraviolet spectrum is recorded between 210 nm and 350 nm for the blank (the background), followed by the sample, and the λ_{max} and absorbance values obtained. From the regression equation formulated earlier, the amount of cocaine in the sample can be calculated.

6.5 Comparison of Cocaine Samples

The profiling and comparison of cocaine is a complex process, the details of which are beyond the scope of this present book. However, it is useful to consider the compounds which might be present in a sample and how they arise.

These include the cocaine content itself, both from the drug and the ecgonines, and other by-products from the process of cocaine synthesis, along with various breakdown products. In addition, there is nor-cocaine, the *N*-demethylated product which arises as a consequence of the use of potassium permanganate to purify the cocaine during the synthetic process, plus the cinnamoyl cocaines and the truxillines, which are congeners of cocaine in the original plant products and which are carried through the synthetic processes used in the manufacture of cocaine. One of the best descriptions of cocaine profiling has been presented by Moore *et al.* [8]. However, such a *complete* analysis as described in this work is unlikely to find routine use because of its complex and highly specialized nature, although it is one of which the drugs chemist should be aware. In this case, illicit cocaine samples, found in a suitcase and wallet, were subjected to chemical derivatization and three distinct gas chromatographic analyses for the detection and relative quantitation of the cocaine manufacturing impurities/by-products. Chemical derivatization of the cocaine samples was accomplished by using heptafluorobutyric anhydride and *N,O*-bis(trimethylsilyl)acetamide and the products then examined by gas chromatography, employing both flame-ionization (GC-FID) and electron-capture (GC-ECD) detection, thus resulting in a detailed comparison of the materials. However, in most laboratories, comparison of the derivatized mixture will suffice.

Summary

Cocaine is derived from the plant *Erythroxylon coca*. This drug is prepared by extraction from the plant of the tropane alkaloids related to cocaine, followed by hydrolysis and further synthetic treatment to give the final product. Both bulk drug and trace materials may be encountered by the forensic scientist. For bulk samples, the analytical sequence includes physical description, presumptive tests, TLC and confirmation by using GC–MS. Interestingly, although the presumptive testing stage may include colour tests, another method can also be employed where the odour is noted following the addition of sodium hydroxide in methanol. Trace samples are analysed by GC–MS of the recovered residues.

Sample comparison involves analysis of a number of the different types of products found in cocaine, usually carried out after an extraction, derivatization and GC–MS protocol.

References

1. Abruzzese, R., 'Coca leaf production in the countries of the Andean subregion', *Bull. Narcotics*, **41**(1), 95–98 (1989).
2. Turner, C. E., Ma, C. Y. and El Sohly, M. A., 'Constituents of *Erythroxylum coca* I: gas chromatographic analysis of cocaine from three locations in Peru', *Bull. Narcotics*, **31**(1), 71–76 (1979).

3. Griffin, W. J. and Lin, G. D., 'Chemotaxonomy and geographical distribution of the tropane alkaloids', *Phytochemistry*, **53**, 623–637 (2000).

4. Fortson, R., *Misuse of Drugs and Drug Trafficking Offences*. 4th Edn, Sweet and Maxwell, London, 2002.

5. Schlesinger, H. L., 'Topics in the chemistry of cocaine', *Bull. Narcotics*, **35**(1), 63–78 (1985).

6. Grant, F. W., Martin, W. C. and Quackenbush, R. W., 'A simple sensitive specific filed test for cocaine based on the recognition of the odour of methyl benzoate as a test product', *Bull. Narcotics*, **27**(2), 33–35 (1975).

7. Gough, T. A. (Ed.), *The Analysis of Drugs of Abuse*, Wiley, Chichester, UK, 1991.

8. Moore, J. M., Meyers, R. P. and Jimenez, M. D., 'The anatomy of a cocaine comparison case – a prosecutorial and chemistry perspective', *J. Forensic Sci.*, **38**, 1305–1325 (1993).

Chapter 7

Products from *Catha edulis* and *Lophophora williamsii*

Learning Objectives

- To be aware of the use of products from *Catha edulis*.
- To understand the identification of products from *Catha edulis*.
- To be aware of the use of products from *Lophophora williamsii*.
- To understand the identification of products from *Lophophora williamsii*.

7.1 Introduction

A number of plants are known to contain drugs which are subject to control. Of the less common seizures that a forensic scientist might encounter are included products of *Catha edulis* Forsk. (Celastraceae), which have the street names 'Khat', 'Cat', 'Qat', 'Jeff' and 'Mulka', and of the peyote cactus, *Lophophora williamsii* (Lem. Ex Salm-Dyck) Coult. (Cactaceae). While not frequently met with, it is important to understand which drugs are present in such materials and how they can be analysed.

7.2 Products of *Catha edulis*

The street drug Khat is derived from the plant *Catha edulis*. The use of this drug (also known as 'Cat', 'Qat', 'Jeff' or 'Mulka') in Western Europe is generally confined to the Somalian and Ethiopian communities [1]. It is also becoming increasingly available and popular in the United States. With continued concern over its use [2], the plant and/or its constituent phytochemicals have come under national control in a number of countries, as well as under international control as a result of the United Nations Convention on Psychotropic Substances, 1971. Cathinone ((*S*)-1-aminopropiophenone) (**1**) is controlled under Schedule I of the Convention, while cathine ((+)-norpseudoephedrine, (*S*,*S*)-norpseudoephedrine) (**2**) is controlled under Schedule III. Both compounds are Class C drugs under the Misuse of Drugs Act, 1971, having been added by the Misuse of Drugs Act, 1971 (Modification) Order, 1986 (S.I. 2230). However, the parent plant material is not controlled in the United Kingdom, although it is controlled in some European countries (Norway and Sweden), the USA and Canada. The plant material contains these drugs at levels of between 0.3–2.0% of the dry weight of the plant.

C. edulis can be cultivated and grown into a small evergreen shrub or tree, about 10 ft tall. Although having originated in Ethiopia, it can also be found growing in Kenya, Malawi, Uganda, Tanzania, the Arabian Peninsula, Zimbabwe, Zambia and South Africa. It is frequently found growing within coffee plantations, between the coffee trees. Closely related species include *Catha spinosa* and *Catha transvaalensis*. While the morphology of *C. edulis* has been described in great detail, it has been concluded by a number of groups that macroscopic and microscopic examination of typical samples likely to be received by a forensic science laboratory is insufficient for the conclusive identification of this material to species and as a consequence, it is necessary to undertake a morphological and phytochemical approach to the identification of this drug [3, 4].

Khat is purchased as bundles of dried and fresh leaves and buds. For transport purposes, it is usually wrapped in plastic bags or banana skins to preserve its moisture content – loss of 'activity' is observed after 48 h if the plant material dries out. However, the drug can also be obtained as dried crushed leaves or in powdered form. These materials are either chewed or used to prepare an infusion which is drunk like tea. The major part of the pharmacological action is due to cathinone, which acts as a central nervous system (CNS) stimulant, promoting excitation, reducing the need to sleep and enhancing communication.

7.2.1 Identification, Quantification and Comparison of Khat Samples

There are few reports in the literature of analyses of material from *C. edulis* and the only known prosecution for the production of cathine and cathinone from Khat (R. vs. Farmer, Lewes Crown Court, 1998) was unsuccessful [5]. Nevertheless, it is still necessary and useful to understand how to identify, quantify and compare such materials.

7.2.1.1 Identification and Quantification of Khat

Basic forensic science principles should be adhered to when carrying out the analysis of a Khat seizure. Prior to any chemical investigation, a full physical description of the material should be made and appropriate samples removed for subsequent analytical work.

DQ 7.1

What details should be reported during a full physical description of a Khat sample?

Answer

A number of details should be recorded. These include information concerning the exhibit (production) label – plus whether or not the chain of custody is complete – and a note of the integrity of the packaging materials. Once removed from the packaging, the materials surrounding the drug sample should be described and, where necessary, removed in a manner which will preserve other evidence features (e.g. fingerprints) which might also be analysed. The form of the drug, and its size, weight, dimensions and smell should all be recorded. If the sample includes plant material, a botanical description might also be useful. For the analysis of such material, leaf and bark (twig) samples should be analysed separately. Powdered forms of the drug should be sampled so as to obtain a representative sample of the material as a whole, and this should then be homogenized prior to analysis.

Wrapping materials should be carefully preserved, since it may be possible to compare the striae on such packaging in attempts to relate various samples to each other.

DQ 7.2

How might the striae of the packaging materials be used to compare samples?

Answer

When rolls of film are made, the processes involved result in striae run-ning along the length of the film. If two packages are produced from the same piece of film, then the pattern of striation might be identical. If two pieces of film are neighbours within the same roll, then there should, in principle, be a physical fit between the striae on these pieces. How-ever, if the striae do not match, this does not preclude the pieces being related – the materials might have been stretched or torn, or be from distant pieces from within the same roll.

Having obtained and described a suitable sample, the next stage is to anal-yse the material for the presence of cathinone and cathine. In addition to these drugs, other structurally similar compounds found in the *C. edulis* plant include norephedrine (**3**). It is well accepted that cathinone is converted into this com-pound by the process of enzymatic reduction. Furthermore, the oxidation prod-uct, 1-phenyl-1,2-propanedione (**4**), and the cathinone dimer, 3,6-dimethyl-2,5-diphenylpyrazine (**5**), may also be identified. These, however, are reported to form as artefacts of the isolation and analytical processes [6].

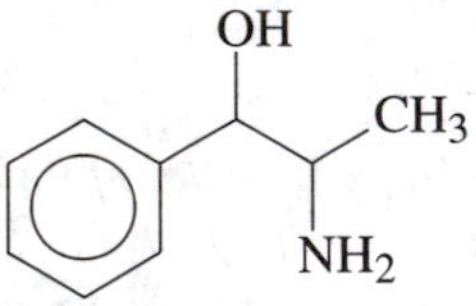

3 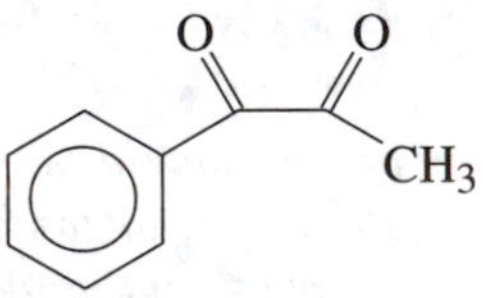   **4**

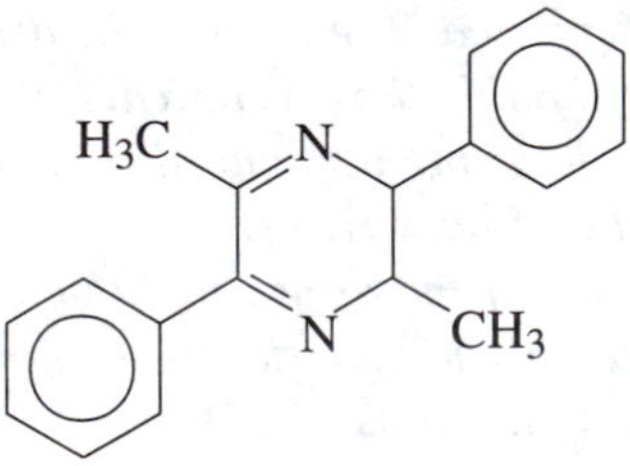

5

In terms of identification of the material as Khat, from *C. edulis*, it is necessary to demonstrate the presence of cathinone, since this is not found in other members of the genus [4].

Typical drug identifications based on chemical analysis depend upon presump-tive (colour) test techniques, followed by chromatographic separation. However, a slightly different approach is required for the analysis of Khat because the spot tests traditionally used for some drugs of abuse (e.g. the Marquis, Simons and Mandelin tests) do not work for cathinone. In general, the material is extracted and the drugs present are determined by using TLC, followed by either HPLC or GC–MS.

In order for chromatographic investigations to be carried out, a representative extract must be obtained which is suitable for the analysis of the drugs. A number of different methods have been employed for extraction of the drug components. All of these are based on the conversion of the drugs to their water-soluble salt forms (for example, as the hydrochlorides) by trituration of the dried plant material in dilute acid (for example, 1 M HCl), basification of the extracts, back-extraction of the drug free bases into an organic solvent which is immiscible with water, drying of the solvent, concentration of the extracts, and then finally analysis of the drug components [6–8], although analysis of simple methanolic extracts has also been reported [4].

DQ 7.3

Why is dilute acid used in the trituration process?

Answer

Under such conditions, the drugs will form water-soluble hydrochloride salts. These can then be selectively extracted into the aqueous acid, thus reducing potential contamination by chlorophylls, carotenoids, fats and other lipophilic materials which are problematic in the analyst of plant materials.

SAQ 7.1

What are the principles behind the basification of the first acid extraction and the subsequent back-extraction?

SAQ 7.2

Why is the organic solvent dried and how might this be achieved?

TLC of Alkaloids from Khat A comparatively simple TLC system can be used for analysis of these drugs, involving separation on silica gel using a mobile phase which causes ion-suppression, and visualizing the chromatogram with ninhydrin [4].

DQ 7.4

Why is ion-suppression required and how might this be achieved?

Answer

Ion-suppression is necessary because of the highly polar nature of the amino groups in these drug compounds. Such groups are readily protonated, forming highly charged species which will sorb strongly on silica gel, thus resulting in considerable tailing during the TLC analysis. The addition of a base to the mobile phase will result in the drugs being present as their free base forms, which will sorb less strongly onto the silica gel and hence reduce the problem of tailing during such analyses.

SAQ 7.3

Why is ninhydrin a suitable reagent for visualization of these compounds on a TLC plate?

The chromatogram is prepared, developed and visualized by following the same principles as those used for other drug classes. While cathinone is reported to separate from the other alkaloids (essential for the identification of Khat), norpseudoephedrine and norephedrine do not separate in such a system and thus further confirmation of the drug identity is required.

HPLC of Alkaloids from Khat HPLC can be used to identify the drugs present in this controlled substance, with resolution of cathinone, norpseudoephedrine and norephedrine being achieved on silica gel (as the stationary phase), with a mobile phase consisting of 1,2-dichloroethane/methanol/acetic acid/diethylamine/water (800:200:10:5:5, by volume), employing UV detection at 257 nm [7]. This method has also been used to quantify the drug components present in these samples. However, retention time data were not provided in this paper.

GC–MS of Alkaloids from Khat Another approach to identifying the drug is to use GC–MS. Two such methods have been described recently. One of these involves components analysis of the underivatized alkaloids on a 30 m HP-5 column (i.d. 0.25 mm; d_f 0.25 μm) under the following conditions: split ratio of 50:1; carrier gas flow rate of 0.87 ml min^{-1}; temperature programme, starting at 40°C, rising to 95°C at a rate of 25°C min^{-1}, holding for 18 min, then rising to 270°C at a rate of 20°C min^{-1}, and holding for 10 min [6]. The retention times of cathinone and cathine were 18 and 19.5 min, respectively. However, it was found in this case that the mass spectra of the two compounds were very similar.

DQ 7.5

Why do the drugs elute in this order?

Answer

The difference in the retention times of the drugs is essentially based upon their boiling points since the molecular weights are very similar.

> *The ketone group of the cathinone will be less polar that the hydroxide*
> *moiety of the cathine. The presence of the latter group will thus cause*
> *this compound to have a higher boiling point and hence in the system*
> *described it elutes second, at the higher column oven temperature.*

Chiral separation of the drugs can be achieved by derivatization of the compounds at room temperature with (R)-(+)-α-methoxy-α-(trifluoromethyl)phenyl acetic acid, followed by separation of the drug derivatives by GC–MS. Racemic mixtures of the drug alkaloids were detected by this technique, hence allowing both identification and comparison of samples [8].

7.2.2 Comparison of Khat Samples

There are a number of difficulties involved with the comparative analysis of Khat samples. At the time of writing,[†] there are no detailed reports in the literature of optimized methods for comparison of such materials. While thin layer chromatographic methods have been used, this technique suffers from lack of resolution. Coupled to the problems associated with artefact formation during the process of analysis, it still remains difficult to achieve a meaningful comparison with any degree of certainty.

7.3 Products of *Lophophora williamsii*

Lophophora williamsii was an important hallucinogen plant to the Ancient Aztecs and even today it is used by the Native American Church, as well as being subject to abuse in other parts of the world. It is a small, button-shaped cactus, known as the peyote, which grows in Mexico and the South-Western United States of America. This cactus rarely grows more than an inch above the ground, with the remainder of the plant being formed by a long underground root. Such cacti can take five to fifteen years to mature. The active compound, mescaline (3,4,5-trimethoxyphenethylamine) (**6**), is a potent hallucinogen. Under United Kingdom legislation, mescaline is a Class A drug under the Misuse of Drugs Act, 1971, although, as with Khat, there is no possession offence related to the intact plant. This contrasts with the United States where both the plant and the active component are classified as Schedule I drugs.

6

[†] May, 2002.

The concentration of mescaline is reported to lie in the range 680–1010 mg per 100 g of fresh plant material [9]. The presence of mescaline allows differentiation from the very similar *Lophophora diffusa*, from which mescaline is absent and pellotine is the dominant alkaloid.

The most commonly encountered form of the drug is 'mescal buttons'. These are the sliced, dried tops of the cactus. Typical dosages are quoted as 100–200 mg of mescaline for a light dose, through to 500–700 mg for a heavy dose. The dose administered depends upon the number of 'buttons' ingested. The onset of effect takes 45–60 min and lasts for 4–8 h. In addition, powdered forms of the drug may also be encountered by the forensic scientist.

In order to determine that a controlled substance is present, it is frequently necessary to carry out a full forensic analysis of such samples. This includes a full physical description, sampling and subsequent chemical analysis, involving presumptive tests, TLC and confirmatory techniques.

7.3.1 Physical Description and Sampling of Materials

Where simple, single package items are to be analysed, the entire production in its wrapping should be described and the drug material then removed from the packaging. It should next be weighed and where possible, a full physical description of the drug sample itself should be recorded. For whole plant specimens, the materials should be dried to constant weight and then cut into two equal pieces, i.e. one for analysis and one for 'court-going' purposes [2].

SAQ 7.4

Why is retention of half of the sample necessary?

The material for analysis, once dried, should be homogenized and if quantitative analyses are to be performed, the materials should then be dried at 110°C to constant weight, prior to analysis.

DQ 7.6

Why should the material be dried to constant weight?

Answer

If the materials are analysed fresh, they will contain different amounts of water, which will affect the reported percentage of mescaline present in the plant. Since the amount of water in each sample will be variable, the results obtained are not directly comparable. If however, the plant material is dried, the analytical data will be comparable, since the amount of water present in the samples can be considered to be nil.

Where a single powdered item is encountered, it should be thoroughly homogenized (using either a blender or the cone-and-square method) and then subjected to the usual analytical sequence, remembering that it is necessary to retain equal portions of material for second and subsequent analyses.

DQ 7.7

Why is it necessary to homogenize plant drug samples?

Answer

Homogenization is needed because whole plants are known to contain different amounts of compounds (drugs) in different tissues within the plant. The amount reported will depend upon the tissue being analysed. Since it is not always easy to determine from the drug sample what the tissue is, or it is not practical to dissect out the different tissues, the easiest way to address this problem is to homogenize the samples. This ensures that any comparison between samples is meaningful.

Where multiple packages have been seized, the following sampling protocol may be adopted. The materials should be divided into groups of indistinguishable materials. Each group should then be counted. For group sizes of less than 10, all of the items should be analysed, for groups between 10 and 100, 10 items should be selected at random and analysed, while for groups of more than 100, the square root of the number should be analysed. Following material selection, the analysis can then proceed.

7.3.2 Presumptive Tests for Mescaline

A number of difficulties are associated with colour tests on fresh and dried plant material, in addition to those associated with analysing a primary amine such as mescaline. These include the fact that the plant material itself may obscure the colour reaction that takes place. Furthermore, a wide range of primary phenethylamines and amphetamines may yield similar colour reactions. The presumptive test which can be satisfactorily used is the Marquis test (see Appendix 1). Positive and negative control tests should also be carried out. An orange–red colour will develop if mescaline is present. However, due to the difficulties associated with interpretation of such results, further confirmation of the presence of mescaline, by using chromatographic and spectroscopic techniques, is required.

7.3.3 TLC Analysis of Mescaline

This is a rapid technique which can be used to determine whether mescaline might be present in a sample [2]. An extract is prepared from the plant material by powdering the latter and then repeatedly extracting this into methanol/880 ammonia (99:1, by volume) at a rate of 4×0.5 ml per 10 mg of plant material. The lipids are then removed by extracting into diethyl ether (4×0.5 ml per 10 mg of plant material) and using the methanol layer following phase separation.

SAQ 7.5

Why is this protocol adopted?

The chromatographic plate is spotted with the extracts, along with the positive and negative controls, where the concentration of drugs in these solutions is of the same order of magnitude. A number of different solvent systems can be employed for this purpose (Table 7.1).

SAQ 7.6

What is the purpose of the ammonia in the mobile phases used in this TLC technique?

The chromatographic plate, following development, is removed from the TLC tank, the solvent front marked, the chromatogram air-dried, and then visualized under UV light (254 nm), after spraying with either ninhydrin or fluorescamine reagents. If ninhydrin is used, the chromatographic plate should be heated at 120°C for at least 15 min to visualize the mescaline. If fluorescamine is used, the plate should be sprayed and heated with a hot-air blower. The developed spots should be observed under UV light (365 nm) where the product formed with mescaline will yield a bright yellow spot, which can be intensified on exposure to ammonia vapour.

7.3.4 HPLC Analysis of Mescaline

Mescaline can be analysed by using a number of HPLC methodologies. Most of these involve reversed-phase systems based upon ion-pairing-type separations [2, 10].

Table 7.1 TLC systems used for the analysis of mescaline

Stationary phase	Mobile phase[a]	R_f[b]
Si gel, dipped in 0.1 M KOH and dried	MeOH/880 NH$_3$ (100:1.5)	0.20
Si gel	CHCl$_3$/MeOH (9:1)	0.10
Si gel	CHCl$_3$/MeOH/880 NH$_3$ (82:17:1)	0.36

[a]Proportions by volume.
[b]Retardation factor of mescaline.

Table 7.2 HPLC operating conditions and parameters used for the quantitative analysis of mescaline [2]

System/parameter	Description/conditions
Column	ODS silica: 15 cm × 4.6 mm i.d.; 3 μm particle size
Mobile phase	Water/MeCN/*o*-phosphoric acid/hexylamine (892:108:5:0.28)[a]
Flow rate	1 ml min^{-1}
Injection volume	5 μl (injection loop)
Detection	UV, at 205 nm

[a]Proportions by volume.

In the first of these systems [2], the dried plant material was powdered and then repeatedly extracted into methanol/880 ammonia (99:1, by volume) at the rate of 4×0.5 ml per 10 mg of plant material. The lipids were removed by extracting into diethyl ether (4×0.5 ml per 10 mg of plant material), and after phase separation, the methanol layer then used for chromatographic analysis. The conditions used for the HPLC separation are presented in Table 7.2. The detection limit for this system is reported to be 500 pg 'on-column'.

DQ 7.8

What is the separation process involved for the system described in Table 7.2?

Answer

In this system, the drug will form an ion-pair with the hexylamine. This has a lipophilic chain which results in good chromatographic perfor-mance. There are two principle mechanisms operating in this situation, in conjunction with each other, resulting in the chromatographic sepa-ration of the components of the mixture. In the first of these, the hexy-lamine and mescaline form an ion-pair in the presence of phosphoric acid. This species is lipophilic and partitions into and out of the stationary phase. In the second mechanism, the hexylamine partitions into the sta-tionary phase, with the mescaline undergoing competitive ion-exchange processes as a result of the presence of both hexylamine and phosphoric acid (in excess) in the mobile phase.

In another methodology, a similar ion-pairing process is involved [10]. How-ever, in this case the sample preparation is described for fresh material. The (fresh) cactus material is pulped and extracted with methanol/ammonia (as in the previous example), or in aqueous buffer at pH 4.0. An ocatdecyl silyl (ODS) sta-tionary phase and a mobile phase of 5.0 mM of aqueous octylamine *o*-phosphate

Table 7.3 GC operating conditions and parameters used for the GC–MS analysis of mescaline [11]

System/parameter	Description/conditions
Column	BP-1: 25 m × 0.22 mm i.d., d_f, 0.33 μm
Injection temperature	240°C
Column oven temperature programme	140°C, no hold; increased to 260°C at 10°C min^{-1}; held for 1 min; increased to 280°C at 20°C min^{-1}; held for 5 min
Carrier gas	He, at a flow rate of 3 ml min^{-1}
Detector	Mass spectrometric[a]

[a] In total ion current (TIC) and selected-ion monitoring (SIM) modes.

were used, employing UV detection at 230 nm. Using this method, it was found that the average amount of mescaline in *L. williamsii* samples was 2.55 mg per g of fresh cactus.

7.3.5 GC–MS Analysis of Mescaline

GC–MS can be used to analyse mescaline in casework samples [11]. In this example, dried, powdered material was soaked in chloroform for 20 min, the mixture centrifuged, and the supernatant recovered and evaporated to dryness. The material was reconstituted in 0.5 ml of methanol and then subjected directly to GC–MS analysis, using the operating conditions given in Table 7.3.

By employing this analytical process, it was possible to determine the presence of mescaline in the casework samples without any ambiguities.

7.3.6 Comparison of Peyote Samples

At the present time,[†] there is no optimized, validated method available for the comparison of mescaline-containing samples. In addition to phytochemical (drug) analyses, another opportunity may exist through the use of DNA analyses (cf. Section 4.4.3.4 above in the case of cannabis profiling). At present, such methods have not been reported in the scientific literature, in the public domain, and thus comparison of peyote samples remains problematic.

Summary

The products obtained from the plant *Catha edulis*, e.g. the street drug Khat, are analysed by using basic forensic science principles. A physical description is first prepared, but, interestingly, the traditional presumptive tests that normally follow this initial stage do not work on the drugs present in such plant material. The

---

[†] May, 2002.

routine process of analysis in this case is therefore continued with TLC, followed by GC–MS, although an HPLC method has been reported. The products of the peyote cactus, *Lophophora williamsii*, are analysed by using a similar sequence, namely physical description, presumptive tests, TLC, GC–MS and quantification employing HPLC. At present, there are no reports in the public-domain literature for comparisons of the drug materials obtained from these two sources, although DNA analysis of the plant materials themselves is theoretically possible.

References

1. King, L. A., 'Drug Classification, including Commercial Drugs' in *Encyclopedia of Forensic Science*, Vol. 2, Siegal, J. A., Saukko, P. J. and Knupfer, G. C. (Eds), Academic Press, New York, 2000, pp. 626–631.
2. United Nations Drug Control Programme, *Recommended Methods for Testing Peyote Cactus (Mescal Buttons)/Mescaline and Psilocybe Mushrooms/Psilocybin*, Manual for use by national narcotics laboratories, United Nations Division of Narcotic Drugs, New York, 1989.
3. Nordal, A. and Lane, M. M., 'Identification of Khat', *Medd. Norsk. Farm. Selsk.*, **40**, 1–18 (1978).
4. Lehmann, T., Geisshusler, S. and Brenneisen, R., 'Rapid TLC identification test for Khat (*Catha edulis*)', *Forensic Sci. Int.*, **45**, 47–51 (1990).
5. King, L. A., *The Misuse of Drugs Act: A Guide for Forensic Scientists*, The Royal Society of Chemistry, Cambridge, UK, 2003.
6. Ripani, L., Schiavone, S. and Garofano, L., 'GC/MS identification of *Catha edulis* stimulant active principles', *Forensic Sci. Int.*, **78**, 39–46 (1996).
7. Schorno, X., Brenneisen, R. and Steinegger, E., 'Analysis of phenylpropylamines from *Catha edulis* using HPLC quantification of the khat-amines for plant material of various origin for plant tissues in various stages of vegetation', *Planta Medica*, **42**, 133–134 (1981).
8. Lebelle, M. J., Lauriault, G. and Lavoie, A., 'Gas chromatographic–mass spectrometric identification of chiral derivatives of the alkaloids of Khat', *Forensic Sci. Int.*, **61**, 53–64 (1993).
9. Helmlin, H. J., Bourquin, D. and Brenneisen, R., 'Determination of phenethylamines in hallucinogenic cactus species by HPLC with photodiode array detection', *J. Chromatogr.*, **623**, 381–385 (1992).
10. Gennaro, M. C., Gioannini, E., Giacosa, D. and Siccardi, D., 'Determination of mescaline in hallucinogenic cactaceae by ion-interaction HPLC', *Anal. Lett.*, **29**, 2399–2409 (1996).
11. Fucci, N. and Chiarotti, M., 1996 'Mescaline in multicoloured statuettes', *Forensic Sci. Int.*, **82**, 165–169 (1996).

Chapter 8

The Analysis of Psilocybin and Psilocin from Fungi

Learning Objectives

- To be aware of the use and control of psilocin- and psilocybin-containing mushrooms.
- To understand the methods available and the difficulties associated with morphological identification of mushroom species within the forensic context.
- To appreciate the methods available for identification and quantification of psilocin and psilocybin in fungal products.
- To be aware of the fact that some drug samples can be identified by using DNA profiling when other methods are not available.

8.1 Introduction

It is now known that there are a large number of fungi which produce hallucinogenic and psychoactive compounds. Of these, the genus *Psilocybe* is the most important [1]. This is a widespread genus, growing from the Arctic to the Tropics, although it is mainly found in temperate regions. Of the 140 species that have been identified, approximately 80 are known to produce hallucinogenic compounds. The most significant of the genus are *P. semilanceata* (FR.) Quel., known as 'liberty caps', and *P. cubensis* (Earle) Singer. The former grows in Middle Europe, North America, Russia and Australia, with the latter being found in Central and South America, Southern Mexico, the West Indies and South-Eastern Asia.

The main alkaloid found in these species is psilocybin (**1**) (0.17–1.07% of dry weight) [2], along with traces of baeocystin (**2**), which is thought to be the biochemical precursor. However, the dephosphorylated, psychoactive drug, psilocin (**3**) is only found in trace amounts (0.11–0.42% of dry weight) [2]. It is thought that *in vivo*, when psilocybin is ingested, it is rapidly converted to psilocin by the action of alkaline phosphatases and that it is this compound which is responsible for the observed psychotropic activity. The amount of drug ingested is highly variable, depending on the species involved, the quality of the material, the habituation of the user and the desired effect. Typically, doses range from 1–5 g of dried material (10–50 g when wet). For psilocybin, the threshold for activity is approximately 2 mg, while a light dose is considered to be 2–4 mg, a medium dose 4–8 mg, a strong dose 8–20 mg and a heavy dose in excess of 20 mg. The onset of activity depends upon a number of factors, but generally occurs between 30–60 min when the material is swallowed (10 min when the material is held in the mouth). The duration of the desired effects is 2–6 h, while the after-effects can last as long as 8 h.

The legislative control of psilocybin-containing mushrooms is complex under the Misuse of Drugs Act, 1971. Both psilocin and psilocybin (the phosphate ester) are classified as Class A drugs. However, cultivation of the mushrooms which contain these compounds and possession of fresh mushrooms is not deemed an offence under the Act. However, if the materials are deliberately 'prepared', by, for example, freezing, drying, cooking, etc., then production of a Class A drug is deemed to have occurred. In the United States, the mushrooms themselves, and the drugs psilocin and psilocybin, are controlled.

8.2 Identification of Psilocybin- and Psilocin-Containing Mushrooms

In order to prove the presence, or otherwise, of these drugs, it is necessary to identify either the controlled substance within the material or the species of fungus involved. The former approach will identify the material as one which contains controlled substances, but not the species involved. The latter route will identify the species and hence through the knowledge of the chemistry of that species, whether it is capable of producing a drug. The methods of identification include using fungal morphology, fungal chemistry and, more recently, the use of DNA profiling. Each of these is described in the following sections.

8.2.1 Identification of Fungal Species from Morphological Characteristics

One means of definitively identifying which species of fungus is present is to make use of morphological characteristics. This is possible provided that each species has one or more unique identifying features. Each identification is based upon a suite of characteristics that leads to a single species. Some examples of the characteristics used for the identification of *Psilocybe semilanceata*, *P. cubensis* and *P. mexicana* are presented in Table 8.1.

Table 8.1 Examples of morphological characteristics used to identify psilocybin-producing species of fungi

Characteristic	Species		
	P. semilanceata	*P. cubensis*	*P. mexicana*
Stipes[a]	White to yellow; 40–70 × 2–3 mm	Hollow; white to yellow; 40–70 × 4–20 mm	Hollow; yellowish pink; 20–60 × 1–2 mm
Cap	Conic to obtuse conic; red brown; width, 10–20 mm; height, 0.5–25 mm	Conic to convex; ochrous to cream to white; width, 25–70 mm	Conic to campanulate; brown to deep ochrous; width, 10–15 mm
Gills[b]	Adnate to–adnexed	Adnate to–adnexed	Sinuate, adnate or adnexed

[a] On the stem.
[b] See Glossary (of Terms) at the end of this text for further descriptions of botanical features.

Although much data has been collated [3], such information illustrates that fungal identification based on fungal morphology requires considerable experience and that it is all too easy to mis-identify specific fungi. It is possible to mis-identify poisonous fungi for those listed here and hence this approach should only really be undertaken by an expert, or under expert guidance.

DQ 8.1

Why is it possible to mis-identify fungal species on the basis of morphology?

Answer

This can occur on account of the large number of characteristics that need to be recognized in order to identify the material to species. There is also a need for considerable experience in this field and an understanding of the technical terms used in the morphological descriptions. Difficulties will be encountered by a novice mycologist because of the overlap in characteristics that will be observed between species, for example, cap size and colour. Added to this is the phenotypic plasticity (the variation in morphology due to the interaction of environment and genetic make-up of the material) which is observed in biological materials and it thus becomes obvious that a great deal of experience is required to achieve this type of identification. The difficulty is compounded if the material is not fresh (and hence sizes and shapes will have changed), or if the material has been 'prepared' or cooked in some way. This is because the most important characteristics used for identification, namely spore size and colour and the basis structures of the cap and gills, may have been lost as a result of such processing.

Nevertheless, if such an approach is to be used, then all underpinning forensic science principles should *also* be followed. The packaging and the data contained thereon should be recorded in a careful manner. The packaging should be opened and the materials subjected to morphological examination. They should then be re-instated and 'signature-sealed' as appropriate.

8.2.2 Identification of Psilocin and Psilocybin Using Chemical Analysis

Due to the potential difficulties involved in using morphological characteristics, chemical analysis is often employed to demonstrate that the material is controlled. However, such an approach will not lead to identification of fungal species.

DQ 8.2

Why will chemical analysis not lead to identification of fungal species?

Answer

*Such identification is not possible because the presence of psilocin and psilocybin is observed in many different species of this genus. Coupled to this, it is also not possible to use the ratios of the amounts of these drugs to identify species because such amounts will vary during the day and between days. Therefore, all that is possible with chemical analysis is to demonstrate the **presence** of the drugs.*

Again, of course, if this approach is used, full forensic science principles should be applied. The packaging should be described and documented and its condition noted. If the analysis is to be quantitative, then the material should be dried to constant weight. Both the fresh weight **and** the dry weight should be recorded.

SAQ 8.1

Why should the material be dried to constant weight prior to quantitative analysis?

If multiple items are to be analysed, then the following protocols (according to the United Nations Drug Control Programme [1]) should be adopted. The materials should be sorted into visually indistinguishable groups and each group treated in the following way. If there are less than 10 packages, they should all be sampled, while if there are between 10 and 100 packages, then 10 should be chosen at random and analysed. If there are more than 100 packages, the square root of the number (raised to the nearest integer) should be analysed.

Following sample selection, it is necessary to homogenize the material. This is because the amount of the drug in different parts of the same specimen and between the specimens themselves is highly variable. In order to be representative, the material must be the 'same all the way through' so that it does not matter where the sample is taken from. Homogenization of the material can be achieved by either grinding in a pestle and mortar, shaking in a plastic bag or grinding in a blender.

SAQ 8.2

What is the principle difficulty associated with the use of a blender for sample homogenization?

Having described all of the materials, selected the samples and homogenized them, the next step in the analysis is to identify the chemical constituents. This is achieved by using presumptive tests, TLC and a confirmatory technique.

8.2.2.1 Presumptive Tests for Psilocin and Psilocybin

Due to the (often) highly coloured nature of fungi, it is advisable to pre-extract the drugs which might be present, in order that colour reaction tests are not obscured by the 'natural' colours of the fungal materials. This can be achieved by finely grinding the material and then suspending a small amount in 1–2 ml of methanol, as appropriate, in a small test-tube. The mixture should be shaken for 5 min to extract the drug(s). A small amount of glass wool (for use as a filter) should be placed above the mixture and the liquid for testing drawn into a clean Pasteur pipette. The colour tests are performed directly on two drops of the extract.

DQ 8.3

If the material is extracted in this way, what is the appropriate negative control and what does it show?

Answer

The appropriate negative control is a sample of methanol treated in exactly the same way as the sample mixture, but without the addition of the fungal material. This shows that the glassware, solvent, glass wool and pipette are free from contamination with compounds which could react in the colour test.

The presumptive tests which are used for these drugs involve use of either the Erhlich or Marquis reagents (see Appendix 1). However, neither test is specific for psilocin and psilocybin. For example, many C2 unsubstituted indoles, including LSD and several uncontrolled substances, will react with Erhlich's reagent, while the Marquis reagent will give colour reactions with many other classes of controlled drugs.

The Use of Erhlich's Reagent This reagent is prepared by dissolving 1 g of *p*-dimethylaminobenzaldehyde in 10 ml of methanol, followed by the addition of 10 ml of orthophosphoric acid. The fungal material is extracted as described above and two drops of the extract are placed on a spotting plate. A negative control and a known sample (positive control) of drug are used in addition to the material being tested. The samples are dried either on a warm heating surface or under a lamp, and then residue redissolved in two drops of Erhlich's reagent. Development of a violet to violet-grey colour indicates the presence of indoles and hence the *potential* presence of psilocin in the sample material.

The Use of Marquis Reagent This reagent is prepared as two components, i.e. (i) eight to ten drops of 40% formaldehyde in 10 ml of glacial acetic acid, and (ii) concentrated sulfuric acid (note that this is slightly different to the system used to test for the presence of amphetamines and opiates). A small amount of material to be tested is placed on a spotting plate, followed by one drop of the formaldehyde solution and two drops of the sulfuric acid. If either psilocin or psilocybin is present, an orange colour will develop.

DQ 8.4

Which other group of drugs develops an orange colour with the Marquis reagent and how may they be differentiated from psilocin or psilocybin?

Answer

Amphetamines containing primary amino groups and unsubstituted benzene rings (amphetamine, methylamphetamine, etc.) also produce an orange colour with the Marquis reagent. They may be differentiated from the drugs being considered here because the amphetamines will not react with Erhlich's reagent.

If, following presumptive testing, it is thought that psilocin and psilocybin might be present in the samples under investigation, then TLC is carried out. This allows a decision as to which samples should be further analysed by using the more expensive instrumental methods and which can be excluded at this stage.

8.2.2.2 TLC of Psilocin and Psilocybin

The fungal materials to be tested should be extracted with methanol as described above. If necessary, the liquid should be subjected to micro-centrifugation to ensure the removal of any micro-particulate material which might interfere with the chromatographic analysis. TLC should be undertaken on silica gel plates impregnated with a fluorescent marker (which fluoresces at 254 nm). The materials being tested, plus the positive and negative controls, should be spotted on the plate, the latter dried and the chromatogram developed in either methanol/ammonia (100:1.5, by volume) or *n*-butanol/acetic acid/water (4:1:5, by volume). When development is complete, the chromatogram should be removed from the TLC tank, the solvent front marked and the plate air-dried at room temperature. The latter should then be visualized in white light, under UV light (at 254 and 360 nm) and after spraying with Erhlich's reagent. At each stage, the colour observed and the R_f value should be recorded. If the colour reactions and the R_f data are the same as those observed for the positive controls, then further confirmatory analysis should be carried out.

8.2.2.3 Confirmation of the Presence of Psilocin and Psilocybin

It is necessary to use instrumental techniques to *confirm* the identity of psilocin and psilocybin if these are thought to be present in the samples under investigation. Of these, GC–MS is the most commonly used method for such confirmatory studies.

GC–MS Analysis It should be noted that psilocybin converts to psilocin by thermal dephosphorylation and hence derivatization of the sample is preferred if psilocybin is to be observed in the mass spectrum. A number of methods for achieving

Table 8.2 GC operating conditions and parameters used for the GC–MS analysis of the trimethylsilyl ethers of psilocin and psilocybin [2]

System/parameter	Description/conditions
Column	HP-5; 30 m × 0.25 mm i.d.; d_f, 0.25 μm
Injection temperature	250°C[a]
Column oven temperature programme	180°C, no hold; increased to 320°C at 20°C min^{-1}; held for 5 min
Carrier gas	He, at a flow rate of 1 ml min^{-1}
Detector	Mass spectrometric in selected-ion mode, monitoring: psilocin, m/z 348, 291 and 290; psilocybin, m/z 485, 455 and 442

[a] Splitless.

this can be found in the scientific literature, but one recently reported concerns the use of N-methyl-N-(trimethylsilyl)-2,2,2-trifluoroacetamide (MSTFA) as the derivatizing reagent [2]. In this paper, it was claimed that the baselines obtained in the spectra were 'cleaner' than if bis(trimethylsilyl)trifluoroacetamide (BSTFA) had been used as the reagent.

In this study, the sample was first lyophilized and 50 mg extracted in 1 ml of $CHCl_3$ for 1 h. The system was then centrifuged at 14 000 rpm for 10 min and filtered through a cotton filter. After evaporation of the supernatant (50 μl) under nitrogen, the residue was redissolved in 30 μl of MSTFA and heated for 30 min at 70°C. GC–MS analysis, using the conditions presented in Table 8.2, was then carried out on 1 μl of the derivatized sample solution.

n-Alkanes can be used as the internal standards. If the retention indices and mass spectra of the positive control and the sample match, then the presence of the respective drug can be confirmed.

DQ 8.5

What is the chemistry of the MSTFA reaction with psilocin?

Answer

MSTFA and psilocin (see structure 3) react to form the following product:

In this case, the derivatization reagent has reacted with the hydroxide group and the secondary amine unit, forming a product which will exhibit much superior chromatographic behaviour when compared to that of the parent drug.

DQ 8.6

Why are the ions at m/z 348, 291 and 290 (see Table 8.2) used for monitoring the diTMS derivative of psilocin?

Answer

Psilocin (M_r, 204) reacts with MSTFA to give the di(TMS) ether species shown above in DQ 8.5. The molecular weight of this product is 348. Therefore, the ion monitored at m/z 348 is the parent ion, $[M]^+$. The ions observed at m/z 291 and 290 represent the ions $[MH - CH_2N(CH_3)_2]^+$ and $[M - CH_2N(CH_3)_2]^+$, respectively.

DQ 8.7

What do the ions used to monitor the tri(TMS) ether derivative of psilocybin (see Table 8.2) represent?

Answer

Psilocybin (see structure 1 (M_r, 284) reacts with MSTFA to give the tri(TMS) ether species, which has a molecular weight of 500. The ions observed in the spectrum can therefore be postulated to arise as follows:

$$m/z\ 485,\ [M - Me]^+$$
$$m/z\ 455,\ [M - (CH_3)_2NH]^+$$
$$m/z\ 442,\ [M - CH_2N(CH_3)_2]^+$$

The method described in this work [2] allows easy and rapid analysis of psilocybin and psilocin in drug samples.

8.2.3 Quantification of Psilocin and Psilocybin by HPLC

HPLC can conveniently be used to quantify psilocin and psilocybin in drug samples. As a *quantitative* method, it has a number of advantages over GC–MS. For example, HPLC does not require any thermal stability of the drug and hence derivatization is not needed. This eliminates the problems associated with ensuring complete derivatization and the risk of sample contamination. Although a number of reported HPLC analytical methods can be found in the literature, we will illustrate the procedure involved by describing a relatively simple method which has been recently reported by Musshoff *et al.* [4].

In this analysis, the fungal material (100 mg) was powdered, extracted into 9 ml of methanol, and then ultrasonicated for 2 h, ensuring that the temperature did not rise above 50°C. The volume of the extract was increased to 10 ml, the mixture centrifuged to remove any solid particulates, and the resultant supernatant used for analysis. The compounds were separated on a Lichrospher 60 RP-Select B column (250 × 4.6 mm i.d., 5 μm particle size), with a solvent flow rate of 1 ml min^{-1}. A binary solvent system was used, comprising 20 mM of KH_2PO_4 as the first solvent (solvent A) and acetonitrile as the second (solvent B), under solvent-gradient conditions which started at 5% of solvent B (held for 2 min), and was then increased to 25% of solvent B after a period of 15 min. The eluant was monitored by using UV detection at a wavelength of 266 nm. Psilocybin and psilocin were found to have retention times of 4.9 and 9 min, respectively.

SAQ 8.3

What is the effect of using the binary solvent system?

SAQ 8.4

What is the principal disadvantage of using gradient HPLC systems for drug analysis?

By using this procedure, it was possible to quantify both psilocin and psilocybin which had been previously identified from GC–MS analysis.

8.3 The Identification of Psilocybin- and Psilocin-Containing Fungi Using DNA Profiling

While it is possible to identify whether or not a material contains psilocin and psilocybin by using instrumental methods, it may be the case that the material has undergone degradation and these drugs are no longer present. It thus might not be possible to confirm such species because of potential damage to the samples under investigation. Under such circumstances, another approach is required. Recent research has shown that it is possible to identify fungal drug species by using DNA profiling [5]. Although such work is still at the experimental stage, it represents an important step forward in developing the array of tests available for identifying drug materials such as these.

Summary

Hallucinogenic fungi belong to a number of different genera. It may be possible to identify the fungal material present in a seizure by using the morphological characteristics (when present) alone, although such evaluation will normally require the involvement of an expert mycologist. Therefore, the traditional route of analysis employed by a forensic scientist, for example, in the case of psilocin- and psilocybin-containing mushrooms of the genus *Psilocybe*, would be physical description, colour testing, TLC and confirmation by GC–MS, with quantification being achieved by HPLC.

However, situations may occur where the morphological characteristics of a fungal material are not clear (and hence species identification is not possible) and/or the samples have been 'damaged', thus leading to decomposition, degradation, etc. of the drug components. Methods have recently been developed, involving DNA profiling, which allow the identification, to species, of psilocin- and psilocybin-producing fungi.

References

1. United Nations Drug Control Programme, *Recommended Methods for Testing Peyote Cactus (Mescal Buttons)/Mescaline and Psilocybe Mushrooms/Psilocybin*, Manual for use by national narcotics laboratories, United Nations Division of Narcotics Drugs, New York, 1989, pp. 20–42.
2. Keller, T., Schneider, A., Regenscheit, P., Dirnhofer, R., Rucker, T., Jaspers, J. and Kisser, W., 'Analysis of psilocybin and psilocin in *Psilocybe subcubensis* Guzman by ion mobility spectrometry and gas chromatography–mass spectrometry', *Forensic Sci. Int.*, **99**, 93–105 (1999).
3. Watling, R. J., 'Hallucinogenic fungi', *J. Forensic Sci. Soc.*, **23**, 53–66 (1983).
4. Musshoff, F., Madea, B. and Beike, J., 'Hallucinogenic mushrooms on the German market – simple instructions for examination and identification', *Forensic Sci. Int.*, **113**, 389–395 (2000).
5. Linacre, A., Cole, M. and Lee, J. C.-I., 'Identifying the presence of "magic mushrooms" by DNA profiling', *Sci. Justice*, **42**, 50–54 (2002).

Chapter 9

The Analysis of Controlled Pharmaceutical Drugs – Barbiturates and Benzodiazepines

Learning Objectives

- To have an appreciation of drugs diverted from licit sources.
- To gain an understanding of the legislative control of barbiturates and benzodiazepines.
- To be aware of the on-column derivatization of barbiturates for GC analysis.
- To understand the identification and quantification techniques used to analyse barbiturates and benzodiazepines.

9.1 Introduction

In addition to drugs from illicit sources, the forensic scientist will sometimes be faced with casework involving samples diverted from legitimate, e.g. commercial, sources. Of these, perhaps the most significant are barbiturates (often found in heroin samples) and benzodiazepines. Both are found under a wide variety of trade names, but the approach to their analysis is broadly similar, either when the material is encountered in the native dosage form, or when it is part of a drug mixture.

Barbituric acid was discovered in the mid-19th century, with the first medical barbiturate, barbitone (**1**), being synthesized in 1903. Phenobarbitone (**2**) was introduced as a pharmaceutical in 1912. Therapeutically, these drugs are used as sedatives, anaesthetics and anticonvulsants. Phenobarbitone is also used in the treatment of epilepsy. During the course of the mid-20th century, increasing knowledge was gained about the side-effects and dependence-related problems associated with barbiturate abuse. At the time of writing,[†] many of the barbiturates are controlled at both the national and international levels. Those met with in the forensic science context are diverted from licit sources and may be encountered mainly as capsules and tablets, injectable solutions and powder forms. They may also be mixed in with other drugs, for example, heroin.

Benzodiazepines, which have the generalized structure shown as **3**, were introduced to replace the barbiturates as tranquilizers, anxiolytics, anticonvulsants and muscle relaxants. All those encountered in the forensic science context have been diverted from licit sources. The majority appear as tablets and capsules, although powders and injectable solutions may also be encountered.

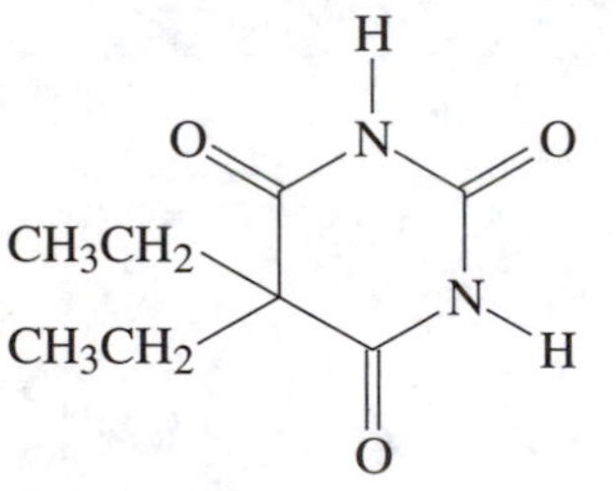

1 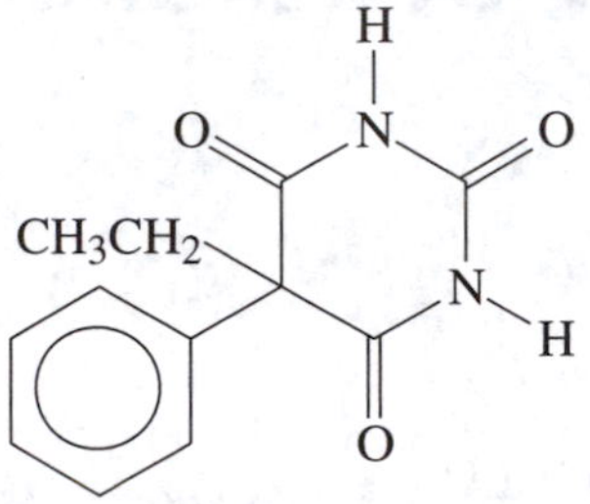   **2**

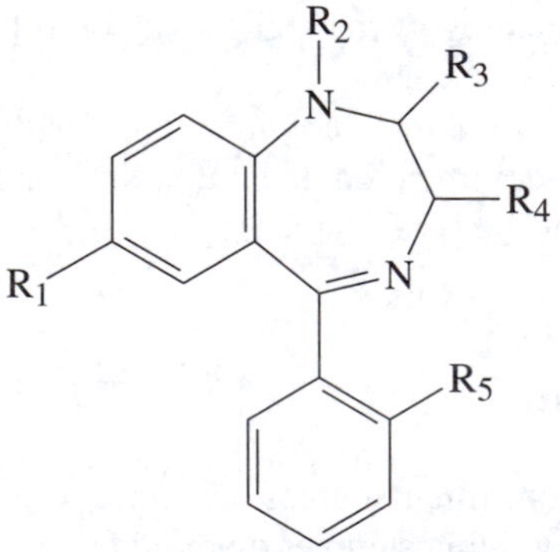

3

Medicinal preparations and dosage forms will occur in carefully controlled doses, e.g. of the order of 2.5–10 mg for benzodiazepines. The dose taken will depend upon the desired effect, the user and a host of other factors.

[†] May, 2002.

Legislatively, 12 of the barbiturates came under international control in 1971, under the UN Convention on Psychotropic Substances. Within the United Kingdom, it was not until 1985 that the barbiturates were generically controlled under the Misuse of Drugs Regulations, 1985, where any 5,5-disubstituted barbituric acids or their stereoisomers, salts and preparations came under control in Schedule 3 of that legislative instrument as Class B drugs. However, the definition only controls the 2-oxo series, and so the 2-thio series, for example, thiopentone (**4**), are excluded from control. The 1,5,5-trisubstituted barbiturate, methylphenobarbitone, does not comply with the generic control and is listed individually as a Class B drug. In addition, quinalbarbitone (secobarbitone) (also 5,5-disubstituted) is listed as a Schedule 2 drug in the Regulations because of its higher toxicity. Under United States legislation, the barbiturates are classified as Schedule II, III or IV drugs, depending upon the particular drug in question.

4

Within the United Kingdom, the benzodiazepines, being individually named, are controlled as Class C drugs, again included by the Misuse of Drugs Regulations, 1985. Under the Misuse of Drugs Regulations, 2001, temazepam and flunitrazepam fall under Schedule 3 control, while the remainder of the benzodiazepines fall under Schedule 4. In the United States, benzodiazepines are controlled as Schedule IV drugs.

9.2 Analysis of Barbiturates and Benzodiazepines

When a seizure of pharmaceutical drugs is made, it may comprise a single dose unit, or many tens or hundreds of thousands of units. The number to be analysed depends upon the legislative system in which the scientist is working, but the following is recommended by the United Nations Drug Control Programme for commercially produced drugs [1, 2]. If between 1 and 50 units are seized, then 50%, to a maximum of 20, chosen at random, should be analysed. Of samples containing between 51 and 100 units, 20 should be analysed. For samples of between 101 and 1000 units, 30 should be chosen, while for samples greater than

1000 units, the square root, rounded to the nearest integer, should be analysed. Once the materials to be examined have been chosen, it is then a matter of identifying and quantifying (where necessary) the specific drug present.

In order to achieve this, a full physical description of the materials should first be carried out. It is sometimes possible, following this process, to feed the information obtained into relevant databases to identify the drug(s) present in the dose form. The *identification* process then becomes a simple matter of *confirmation*. If the dose form is not included in the databases, however, a full chemical analysis, including drug extraction from the tabletted material, presumptive testing, thin layer chromatography and a confirmatory technique must be undertaken.

9.2.1 Extraction of Barbiturates and Benzodiazepines from Dose Forms

Barbiturates in either the free acid or salt forms are readily soluble in methanol and thus this is the solvent of choice for extraction in *qualitative* analysis. A known mass of the dose form is dissolved in a volume of methanol to provide the drug component at a working concentration of between 1 and 20 mg ml^{-1}. The extract should be filtered or centrifuged prior to analysis in order to remove any unwanted particulate materials. For *quantitative* analysis, ethyl acetate can be used as the extraction solvent – if the drug is in the free acid form – with the extract treated in the same way as described above. If the original material is in the salt form, then the drug can be converted to the free acid form and extracted if required.

Benzodiazepines may conveniently be extracted into methanol for *both* qualitative and quantitative analyses. The dose form should be triturated in methanol and as with barbiturates any solid material removed by centrifugation or filtration prior to analysis of the drug in solution.

For both types of drug, presumptive tests, TLC and confirmatory analysis is then carried out on the prepared extract.

9.2.2 Presumptive Tests for Barbiturates and Benzodiazepines

Several presumptive tests are available for barbiturates and benzodiazepines. These are not as 'general' as those used for other drug classes (for example, the Marquis test for opiates and amphetamines (including ring-substituted species)), but have the disadvantage that they do not discriminate between the drugs within the specific class. These tests are described in the following sections.

9.2.2.1 Dille–Koppanyi Test for Barbiturates

This colour test involves the use of two reagents. The first of these is a solution of 0.1 g of cobalt acetate and 0.2 ml of glacial acetic acid in 100 ml of methanol, with the second comprising 5 ml of isopropylamine in 95 ml of methanol. To test materials for the presence of barbiturates, three drops of the first reagent, followed by three drops of the second, are placed on the test material contained in the well of a spotting plate (positive and negative controls should also be undertaken). A purple colour develops if barbiturates are present in the test sample. This test is convenient because there are very few other substances which give such a colour reaction. However, it does not discriminate between specific barbiturates and thus further analyses are required.

9.2.2.2 Zimmerman Test for Benzodiazepines

Again, two reagents are required for this test. The first of these is a 1% (wt/vol) solution of 2,4-dinitrobenzene in methanol, with the second being a 15% aqueous solution of KOH. One drop of the first reagent, followed by one drop of the second, is added to the test substrate. The development of a red–pink colour putatively indicates the presence of diazepam or some other benzodiazepine. However, it should be noted that a number of other materials will also result in a similar colour reaction.

9.2.3 TLC of Barbiturates and Benzodiazepines

Having carried out presumptive tests for these drugs, if a positive result occurs, then further analysis is required. The next phase in the analysis is TLC. Methanol solutions of the drugs can be used, while the standards employed should include the drug which has been suggested through physical examination of the dosage form. However, these drug groups also present some special difficulties in their TLC analysis – these are discussed below.

9.2.3.1 TLC Analysis of Barbiturates

In this case, the TLC system most commonly employed uses silica gel plates and a mobile phase of ethyl acetate/methanol/25% ammonia (85:10:5, by volume). The plates are prepared and the chromatogram developed in the standard way. After development, the plate is removed from the mobile phase, the solvent front marked, and the plate dried. Visualization of barbiturates is best achieved by the use of a mercuric chloride–diphenylcarbazone reagent. The latter is prepared as two component solutions, i.e. (i) 0.1 g of diphenylcarbazone in 50 ml of methanol, and (ii) 0.1 g of mercuric chloride in 50 ml of ethanol. These solutions should be freshly prepared and mixed just before use. The presence of barbiturates will give rise to blue–violet spots on a pink background when using this reagent system.

DQ 9.1

What are the health and safety issues associated with the use of this (spray) reagent?

Answer

All mercuric compounds are extremely toxic, and thus there are health hazards associated with the use of the mercuric chloride–diphenyl carbazone reagent. Testing should ideally be carried out in a fume cupboard, with standard safety procedures being observed, e.g. the use of rubber gloves, etc. In addition, there is the risk of potential contamination of the working area – this should be thoroughly cleaned after use. Finally, special precautions need to made with respect to the disposal of any waste materials after testing has been carried out.

DQ 9.2

Table 9.1 below shows the data obtained for a number of different barbiturates after TLC analysis under the conditions described above. Why are the R_f values presented for the alkyl compounds so similar and what is the difficulty with this fact?

Answer

The reason that the R_f data are so similar is that straight-phase TLC is principally based on sorption–desorption processes. The only variation between the barbiturates is the nature of the substituent at the C5 position. For those compounds with alkyl chains, for example, barbitone, butobarbitone and pentobarbitone, the length and polarity of the side-groups will affect the overall polarity of the molecule through van de Waals forces, which are relatively weak in comparison to other types of interactions which result from polarity considerations. The consequence of this is that there is very little difference in polarity to exploit for the chromatographic separation of these molecules. (Similar polarization effects are also operational in the case of the ring-substituted compounds – see below.) When these facts are coupled to the lack of resolving power of TLC in general, it becomes clear that there is insufficient difference to identify barbiturates on a definitive basis by using this chromatographic technique.

SAQ 9.1

Why is the R_f value of phenobarbitone less than cyclobarbitone (see Table 9.1)?

9.2.3.2 TLC Analysis of Benzodiazepines

The benzodiazepines are a diverse group of drugs and require a combination of different TLC solvent systems to resolve the many drugs in this group. The systems which have been used include chloroform/acetone (80:20 and 90:10, by volume) and cyclohexane/toluene/diethylamine (75:15:10, by volume). The

Table 9.1 Structures of a number of different barbiturates and corresponding chromatographic data obtained after analysis by TLC (DQ 9.2 and SAQ 9.1)

Barbiturate	Structure	R_f
Barbitone		0.33
Butobarbitone		0.39
Pentobarbitone		0.44
Phenobarbitone		0.29
Cyclobarbitone		0.35

chromatographic plates are prepared, developed and dried in the same way as that used for barbiturates. However, the principle difficulty in this case lies in the fact that there is no *specific* development reagent for benzodiazepines. The reagents that have been used include 1 M H_2SO_4, which is sprayed onto the plate, and after heating is subsequently viewed under UV light (at 366 nm). Any fluorescent spots are recorded. The chromatogram is then oversprayed with acidified potassium iodoplatinate reagent, which forms purple spots. If the colour reactions observed for the samples and standards match, then further analysis is required.

9.2.4 Confirmatory Analysis of Barbiturates and Benzodiazepines

Both barbiturates and benzodiazepines can be identified by using GC–MS methodologies, although each drug class requires a different pre-treatment routine prior to analysis. It should also be remembered, as with all chromatographic analyses, that blanks should be run between each sample under investigation and check standards analysed as required by the quality assurance (QA) procedures in place in the laboratory.

9.2.4.1 GC–MS Analysis of Barbiturates

Due to their highly polar nature, barbiturates require derivatization prior to analysis by GC–MS. In such cases, the sample is dissolved in methanol, centrifuged, the supernatant recovered and placed in a derivatizing vial and is then 'blown down' under nitrogen. The derivatization procedure, using 0.2 M trimethylanilinium hydroxide in methanol, is, in principle, the same as that used for other pre-column derivatizations. However, with this system the reaction does not occur immediately because insufficient activation energy is available at ambient temperature for this to take place. Direct transfer of the reaction mixture onto the heated injection block of the gas chromatograph overcomes this problem and the derivatization reaction can then proceed. Such a reaction, an example of 'flash alkylation' is illustrated in Scheme 9.1.

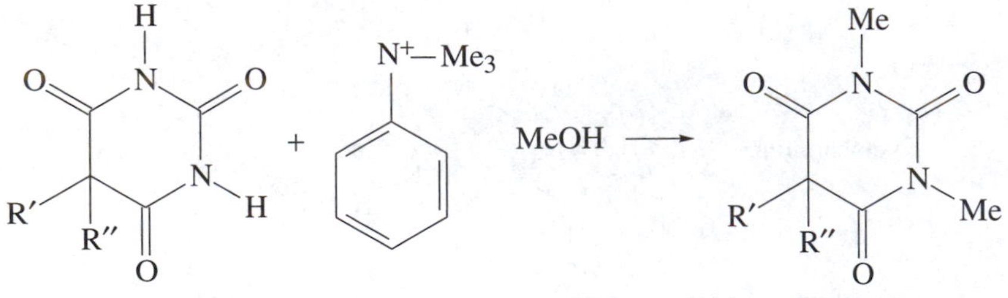

Scheme 9.1 *N*-methylation of a generalized barbiturate.

Table 9.2 GC operating conditions and parameters used for the GC−MS analysis of barbiturate derivatives

System/parameter	Description/conditions
Column	BP-1: 25 m × 0.22 mm i.d.; d_f, 0.5 μm
Injection temperature	290°C
Column oven temperature programme	200°C, no hold; increased to 260°C at 4°C min^{-1}
Carrier gas	He, at a flow rate of 1 ml min^{-1}
Split ratio	20:1
Detector	Mass spectrometric, temperature and settings as required

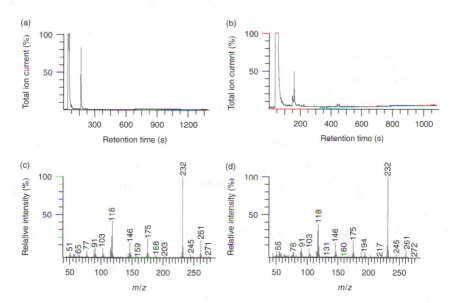

Figure 9.1 GC−MS data obtained for laboratory standard and casework samples of phenobarbitone, derivatized with trimethylanilinium hydroxide: (a) gas chromatogram of standard (t_r, 166 s); (b) gas chromatogram of casework sample (t_r, 164 s); (c) mass spectrum of standard; (d) mass spectrum of casework sample.

Having derivatized the sample, it can then be analysed by GC−MS, using the operating conditions shown in Table 9.2.

By employing this system, it is possible to analyse and identify barbiturates in casework samples, as illustrated in Figure 9.1 for the analysis of phenobarbitone. If the retention time and mass spectral data obtained for the standard and the components of the drugs sample are the same (as shown in this case), then an identification can be called.

DQ 9.3

The mass spectrum of the *N*-methylated derivative of phenobarbitone (**5**) is shown below in Figure 9.2. Explain this spectrum as fully as you can, considering the major ions with $m/z > 110$.

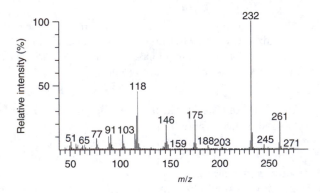

5

Answer

The major ions may arise as follows:

m/z 261,	$[M + H]^+$
m/z 245,	$[M - CH_3]^+$
m/z 232,	$[M - 2CH_3 + 2H]^+$
m/z 175,	$[M - CONMeCO]^+$
m/z 146,	$[M - CONMeCONMe]^+$
m/z 118,	$[M - CONMeCONMeCO]^+$

9.2.4.2 GC–MS Analysis of Benzodiazepines

Benzodiazepines can be conveniently identified by GC–MS analysis. This class of compounds does not require derivatization and thus can be analysed directly

Figure 9.2 Mass spectrum of the *N*-methylated derivative of phenobarbitone (DQ 9.3).

Table 9.3 GC operating conditions and parameters used for the GC–MS analysis of benzodiazepines

System/parameter	Description/conditions
Column	BP-1: 25 m × 0.22 mm i.d.; d_f, 0.25 μm
Injection temperature	275°C
Column oven temperature	250°C[a]
Carrier gas	He, at a flow rate of 1 ml min^{-1}
Split ratio	20:1
Detector	Mass spectrometric, temperature and settings as required

[a] Isothermal.

from a solid-free methanol extract of the drug sample, by using the (typical) operating conditions shown in Table 9.3.

Again, if the retention time and mass spectral data obtained for the standard and the casework sample are the same, then an identification can be called.

9.2.5 Quantification of Barbiturates and Benzodiazepines

In some cases, the forensic scientist will be required to quantify barbiturates and benzodiazepines in drug samples. Due to the problems associated with sample derivatization, quantification of barbiturates can best be achieved by using HPLC. Benzodiazepines are thermally labile compounds and thus HPLC is again the method of choice for qualitative analysis.

9.2.5.1 Quantification of Barbiturates by HPLC

Barbiturates can be prepared for HPLC analysis by dissolution of the drug sample in methanol at a chosen concentration, followed by removal of any solid particulate material by filtration, etc. A typical set of HPLC operating conditions used for the analysis of barbiturates is shown in Table 9.4.

Table 9.4 HPLC operating conditions and parameters used for the analysis of barbiturates

System/parameter	Description/conditions
Column	Spherisorb ODS-2: 25 cm × 4.6 mm i.d.; 5 μm particle size
Mobile phase	40% acetonitrile in water
Flow rate	1 ml min^{-1}
Injection volume	5–10 μl[a]
Detection	UV, at 235 nm

[a] Fixed loop size.

Such a system can be used to quantify barbiturates with relatively short alkyl substituents at the C5 position. The analytes will separate, eluting in order of lipophilicity, e.g. barbitone, butobarbitone, pentobarbitone, etc. However, as the substituents become increasingly apolar as the carbon chain-length increases, so the percentage of the acetonitrile in the mobile phase must also be increased.

SAQ 9.2

Why is it necessary to increase the proportion of acetonitrile in the mobile phase as the alkyl chain-length of the C5 substituents increases?

DQ 9.4

Why is an isocratic mobile phase system preferred to a gradient system in the HPLC analysis of barbiturates?

Answer

Isocratic systems require no long equilibration times and thus are preferred in such analysis.

DQ 9.5

Why is UV detection at 230 nm used?

Answer

Barbiturate compounds do not contain good chromophoric groups. As a consequence, short wavelengths must be used for the detection of such drug components. This is also the reason why diode-array (UV) detection does not provide a very discriminating means of identifying barbiturates after separation by HPLC.

Quantification of barbiturates in thus achieved in the usual way by employing this approach.

9.2.5.2 Quantification of Benzodiazepines by HPLC

The basic principles behind the quantification of benzodiazepines are the same as those applicable to barbiturates. A typical set of HPLC operating conditions used for such analyses are shown in Table 9.5. Again, the compounds elute in order of increasing lipophilicity and, owing to the lack of any good chromophoric groups (cf. the barbiturates), a UV detection wavelength of 240 nm is also used for such materials.

While it is sometimes possible to identify and confirm the drug content of these pharmaceutical materials through research into various internationally available databases and other resources, it is still necessary within the forensic science

Table 9.5 HPLC operating conditions and parameters used for the analysis of benzodiazepines

System/parameter	Description/conditions
Column	Spherisorb ODS-2: 25 cm × 4.6 mm i.d.; 5 μm particle size
Mobile phase	MeOH/water/0.1 M phosphate buffer at pH 7.25 (55:25:20), or (70:10:20)^a
Flow rate	1.5 ml min^{-1}
Injection volume	5–10 μl^b
Detection	UV, at 240 nm

^{a} Proportions by volume.
^{b} Fixed loop size.

context to prove the identity of the drug experimentally and where necessary, quantify the component. Drugs derived from commercial sources are treated in exactly the same way as those obtained from illicit sources and the same rigorous forensic science principles should always be applied in their evaluation.

Summary

Barbiturates and benzodiazepines are usually encountered by the forensic scientist as tablet or capsule preparations that have been diverted from licit sources, or, particularly in the case of barbiturates, as 'cutting agents' in other drug materials (for example, phenobarbitone in heroin samples).

In some cases, identification and confirmation of the dose form can be achieved by using internationally available databases. If this is not possible, then the traditional process of physical description, presumptive testing, TLC and GC–MS should be followed to identify the drug components. However, some benzodiazepines are thermally labile and in such cases HPLC, possibly with diode-array detection, is often the chosen method of analysis. The latter technique is, in addition, the preferred method for quantification purposes.

Such drugs, when obtained from licit sources, are very pure and it is therefore particularly difficult, if not impossible, to compare the samples in order to determine if they once originated from the same batch.

References

1. United Nations Drug Control Programme, *Recommended Methods for Testing Benzodiazepine Derivatives under International Control*, Manual for use by national narcotics laboratories, United Nations Division of Narcotic Drugs, New York, 1988.
2. United Nations Drug Control Programme, *Recommended Methods for Testing Barbiturate Derivatives under International Control*, Manual for use by national narcotics laboratories, United Nations Division of Narcotic Drugs, New York, 1989.

Chapter 10

Current Status, Summary and Conclusions

Learning Objectives

- To understand the current status of drugs analysis.
- To appreciate the general problems associated with drug profiling.
- To know the more frequently reported research methods for drugs analysis.
- To be aware of the opportunities which exist for a drugs analyst.

10.1 Current Status of Drug Analysis

At the present time, it is possible to carry out identification and quantification of a wide variety of drugs, ranging from those which are entirely herbal or fungal in origin (*Cannabis* and its products), through those which are semi-synthetic (cocaine and diamorphine), to those which are entirely synthetic (the amphetamines). A wide variety of techniques can be applied for their analysis and it is rare that an issue of sensitivity becomes apparent. In terms of drug identification and quantification, the drug analyst is in a particularly strong position.

However, the position changes with respect to the comparison of drug samples. Such comparison (profiling) procedures can be considered to consist of five stages, as follows:

- sample selection
- sample homogenization
- sample preparation for analysis
- sample analysis
- data interpretation and reporting

Even with the (apparently) simple task of sample selection, we are presented with a great number of difficulties which still remain to be resolved. These include the following:

- How many samples should be taken and what proportion of the whole should these represent?
- Should it be assumed that there is one drug, two drugs or more than this present in a sample, or that the accused is innocent and that there are no drugs present?

Each of the above aspects have been studied in great detail and a number of mathematical models proposed for evaluation studies, but there is currently no universally adopted method (apart from the United Nations recommendations).[†] However, the latter themselves present their own problems – how are random numbers assigned to individual doses in a batch of thousands so that the samples can be chosen truly randomly?

A further question that might be asked is how should drugs be homogenized once selected? Large batches may be treated by using blenders or grinders, but in this case there is the associated difficulty of ensuring that the equipment is clean before use. Small samples can be handled by the cone-and-square method, but what of medium-sized samples which may be too large for the latter technique but too small for the former?

Much is known about sample preparation for comparison purposes in cases of the more commonly encountered materials, for example, heroin, and *Cannabis* and its products. However, considerably less information is available concerning the stabilities of the newer 'designer' drugs in the solvents used to prepare them, what the extraction efficiencies are for the different impurities, and whether, for example, artefacts are formed as a consequence of a particular preparation process. Research is clearly needed in these areas and this has only really just begun.

Another issue is whether or not the analytical techniques themselves are optimized. Have the methods used been fully scrutinized? Are they valid for the synthetic drugs made by the routes encountered today, plus are they separating the anticipated impurities without the expected artefacts?

There is also the issue of which numerical method should be used for drug comparison investigations. This has been well studied for heroin, but the arena is wide open for analysis and numerical comparison of *Cannabis* and its products, cocaine, amphetamines, tryptamines and other synthetic or semi-synthetic drugs. How these methods should be reported has still not been fully explored.

There are also a number of new analytical techniques available for the comparison of samples. Traditionally, drugs are compared on the basis of their impurity content. Little attention, until recently, had been applied to the use of DNA profiling for drug identification and comparison – such an approach offer great

---

[†] As at May, 2002.

potential for application in this area. A wide variety of alternative methods, involving the analysis of e.g. metal ion concentration, stable isotope ratios and occluded solvent content, to name just a few, have been reported in the scientific literature, although these have not yet found widespread application in case-work studies. The question must be raised – are these approaches worth further investigation?

10.2 Conclusions

Drug analysis is an exciting and rewarding field. The analyst is in a strong position to identify and quantify drugs and on occasion, may encounter significant challenges when having to compare drugs. It also provides an opportunity to learn and understand some fascinating chemistry. While many drug cases are mundane, occasionally an interesting or problematic case appears which challenges the ability of even the best and most experienced drugs chemist. This is especially true in drugs comparison cases. With the ever increasing number of synthetic drugs now available 'on the street', there is a need to develop new methods of analysis, especially for drug comparison, either based upon the existing analytical methodologies, or indeed, for new approaches to be developed. As can be seen from the above, this field of analytical science offers a very wide range of opportunities for further research and development.

Finally, it is hoped that study and application of the principles outlined in this text will allow the reader, whether pursuing an undergraduate or postgraduate course or newly in post as a forensic science practitioner, to progress and develop into a successful drug analyst.

Summary

While it is possible to carry out the necessary processes of identification, quantification and comparison on a wide range of various drug samples and drug components in different materials, a number of questions and problems still remain. These include the need to address how samples should be taken, the processes to be adopted for homogenization once selected and the requirement to develop new optimized methods and techniques for sample comparison purposes, either based upon existing analytical methodologies or involving new technologies. The wide range of opportunities for research and development in this field presents the drugs analyst with an exciting challenge for the future.

Appendices

1 Presumptive (Colour) Tests

There is a wide range of colour tests, involving a number of different reagents, available to the drug chemist for presumptive test purposes. The following gives details of such reagents and the procedures adopted for those tests which are most frequently used. However, it should be remembered that both positive and negative controls should always be incorporated into the experimental protocol being adopted.

Cobalt Isothiocyanate

This is used as a 2% (wt/vol) solution in water, and added directly to the test substrate. A blue coloration is produced with cocaine and its congeners, diamorphine (but no other opiates) and Temazepam (but no other benzodiazepines).

Dille–Koppanyi Reagent (for Barbiturates)

The components of this reagent consist of 0.1 g of cobalt acetate in 10 ml of 0.2% glacial acetic acid in methanol (1) and 5% isopropylamine in methanol (2). Testing is carried out by adding a small amount of component (1) to the test substrate, followed by a few drops of component (2). A blue colour will develop if barbiturates are present in the sample.

Duquenois–Levine Reagent (for Cannabis Products)

The components of this reagent are 2.5 ml of acetaldehyde and 2 g of vanillin dissolved in 100 ml of 95% ethanol (1), concentrated hydrochloric acid (2) and chloroform (3).

Testing is carried out as follows. Five drops of solution (1) are added to the material under investigation, which has previously had a drop of ethanol added to it to solubilize the cannabinoids, and the system shaken. Five drops of solution (2) are then added and after shaking again, 10 drops of solution (3) are added. The whole is then shaken thoroughly and the two phases allowed to separate. Cannabinoids rapidly form a blue complex which extracts into the **chloroform** layer. Beware, however of 'false positives' that can be confused with reactions from 'old' *Cannabis* products.

Ehrlich's Reagent[†] (for Indole Alkaloids)

This reagent is prepared by dissolving 1 g of *p*-dimethylaminobenzaldehyde in 10 ml of methanol and adding 10 ml of concentrated phosphoric acid. The reagent is then added directly to the test substrate. A grey/violet coloration is produced with those compounds containing the indole alkaloid nucleus (e.g. LSD and psilocin).

Mandelin Reagent

This reagent is prepared as a 1% solution of ammonium metavanadate in concentrated sulfuric acid, which is then added directly to the test substrate. Orange colour reactions are observed with cocaine and related compounds, with olive green colours being produced if opiates are present.

Marquis Reagent

This is prepared by dissolving 5 ml of 40% formaldehyde in 100 ml of concentrated sulfuric acid. The reagent is then added directly to the test substrate. A purple–olive coloration is produced with opiates, and yellow–orange with amphetamine, methylamphetamine, psilocybin and psilocin, among others. A grey-purple colour reaction is obtained with ring-substituted amphetamines.

Scott Reagent (for Cocaine)

This is a modification of the previous test, with the reagent consisting of a 2% (wt/vol) solution of cobalt isothiocyanate in water, diluted with an equal volume of glycerine (1), concentrated hydrochloric acid (2) and chloroform (3). Testing for cocaine is carried out as follows. A small amount of the material is placed in a test-tube and five drops of component (1) added, when a blue colour will develop if cocaine is present. One drop of concentrated hydrochloric acid (2) is then added and the blue colour, if this has resulted from the presence of cocaine, should disappear, leaving a pink solution. Further confirmation is provided by the addition of several drops of chloroform (3), whereupon an intense blue colour

---

[†] Also known as Van Urk's reagent.

will develop in the lower (chloroform) layer if cocaine is present in the original sample material.

Zimmerman Reagent (for Benzodiazepines)

This reagent consists of two components solutions, i.e. 1% 2,4-dinitrobenzene in methanol (1) and 15% aqueous KOH (2). To perform this test, a few drops of component (1), followed by a few drops of component (2), are added directly to the test substrate. The rapid development of either a red–purple or pink colour indicates the presence of benzodiazepines in the substrate.

2 Less-Common Controlled Substances

In addition to the major classes and types of drugs described in the main body of this text, there are a number of other controlled substances which may be less frequently encountered by the forensic scientist. Some examples of these are presented in the following table. (Note that this selection is not intended to be exhaustive – merely illustrative.)

Name	Structure	Dose/form of administration	Desired effect
PCP, phencyclidine		5–10 mg Most commonly as cigarettes dipped in a solution of the drug	Dissociative
2C-B (4-bromo-2,5-dimethoxy-phenethylamine)		Low, 5–15 mg; medium, 10–25 mg; high, 20–50 mg Usually taken orally	Hallucinogenic
U4Euh, Euphoria, 4-methylaminorex (4-methyl-5-phenyl-2-aminooxazoline)		Low, 5–15 mg; medium, 15–20 mg; high, >20 mg Usually taken orally or smoked	Stimulant

(*continued overleaf*)

Name	Structure	Dose/form of administration	Desired effect
2C-T-7 (2,5-dimethoxy-4-(*n*)-propylthio-phenethylamine)		Low, 10–20 mg; medium, 15–30 mg; high, 10–50 mg Usually taken orally or insufflated[a]	Hallucinogenic
Ketamine(2-(2-chlorophenyl)-2-(methylamino)-cyclohexanone)		Low, 15–30 mg; medium, 30–75 mg; high, 60–125 mg Usually taken orally or insufflated[a]	Dissociative hallucinogenic

[a]Ingested via the nose.

Responses to Self-Assessment Questions

Chapter 2

Response 2.1

There are several advantages. These include the possibility of establishing a link in terms of tablet manufacturing, since a tabletting operation may be supplied with drugs from more than one drug source – the drugs may not necessarily be produced and tabletted on the same site. In addition, tablets with the same ballistics features may contain different drugs, but such features may help to establish a link.

Response 2.2

The reason for this is that in solution the drug may decompose faster, e.g. through hydrolysis, oxidation, polymerization or other artefact-forming processes. Drying the swab greatly reduces this risk, thus allowing for an increase in the possible time-period between the swab being collected and subsequently analysed.

Response 2.3

This is because it may be necessary to compare samples at a later date, especially if 'tablet databases' have been established. If the ballistics features are destroyed during the sampling process, they cannot subsequently be recovered.

Response 2.4

Methanol is used because it satisfies a number of criteria which represent a 'good solvent' for this drug class. In particular, this solvent will freely dissolve amphetamines (and at this stage it is not yet known *which* species is present), is volatile (hence allowing rapid and easy preparation of the chromatogram prior to development) and easy to handle, as well as being unreactive towards these drug types.

Response 2.5

Methanol itself is relatively polar and, in addition, traces of water will always be found in methanol solutions. These will lead to hydrolysis and decomposition of the amphetamines in the sample under investigation. Therefore, all solutions should be freshly prepared immediately before analysis.

Response 2.6

This is necessary in order to convert the amphetamine from the sulfate form of the drug to its free base form, and hence to improve the efficiency of the reaction with carbon disulfide. Such a conversion is achieved by the formation of ammonium sulfate and amphetamine, since the ammonia neutralizes the acid. Since ammonia is a strong base, it reacts with the sulfuric acid irreversibly, thus rendering the re-formation of amphetamine sulfate impossible.

Response 2.7

The fragments have the following origins:

m/z 178, M^+
m/z 135, $[M - CS]^+$
m/z 119, $[M - NCS]^+$
m/z 86, $[M - C_7H_7]^+$

Response 2.8

The structures of these derivatives are as follows:

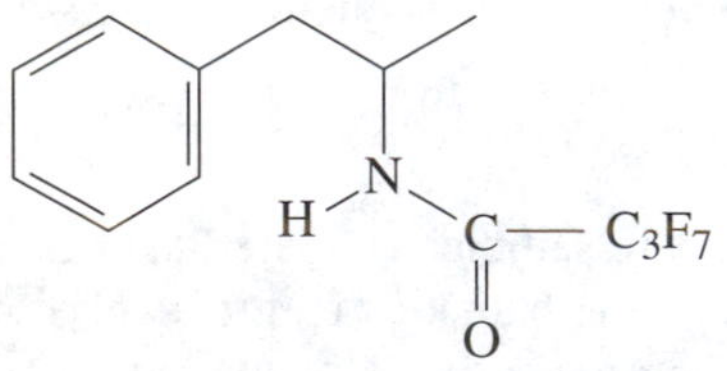

6 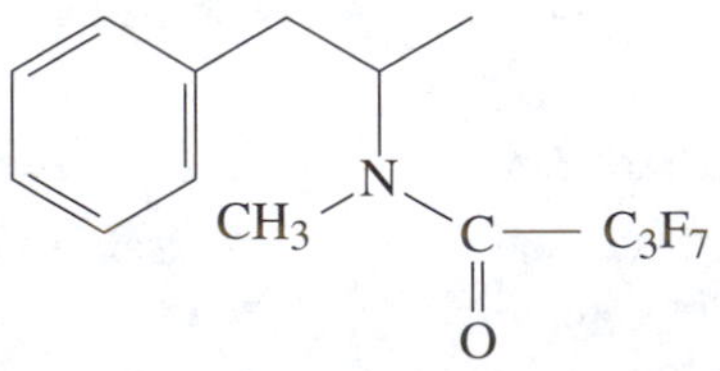   7

The ion at m/z 254 arises from the loss of the C_7H_7 unit from the parent ion, i.e. $[M - C_7H_7]^+$.

Response 2.9

This is to allow all of the active sites on the silica gel to acquire the correct charge and thus bind to the ions from the mobile phase. Where ion-exchange processes are used for HPLC separations, it is particularly important that the stationary phase is fully equilibrated with the mobile phase so that reproducible and repeatable results can be obtained. If the stationary phase is not fully equilibrated, the separation processes for each analysis will be different.

Response 2.10

This is to reduce the risk of 'column priming', which, if allowed to occur, can take a considerable period of time to resolve. Column priming is a phenomenon which occurs when samples at high concentrations are being analysed, with the analyte becoming bound to the stationary phase, only to be eventually eluted at some later time.

Response 2.11

These are samples of known identity and concentration which are used to determine that the chromatographic system and detector are functioning correctly. This is achieved through examination of the retention time and, sometimes, a specific chromatographic property, for example, peak asymmetry, area or height. If the system is performing correctly, then the results obtained from such check standards should be the same every time.

Chapter 3

Response 3.1

This is necessary since the presence of any solid material on a TLC plate will lead to tailing in the chromatogram. In a chromatographic instrument, e.g. an HPLC system, such material will simply block the chromatographic pathway, thus rendering the data worthless and potentially damaging the instrument.

Response 3.2

The dilute acid results in the formation of the charged, salt form of the amide. This is because the proton from the acid hydrogen-bonds to the lone-pair of electrons on the nitrogen atom, with the latter then becoming positively charged, and hence balanced by the anion of the acid. Such polar molecules are freely water-soluble and are easily and efficiently dissolved. The chloroform is used to remove any apolar material before basification, while addition of the sodium bicarbonate results in the neutralization of the acid and formation of the free base form of the

LSD. The latter will then extract preferentially in the chloroform into which the extraction is effected. Three washes are used to improve the extraction efficiency.

Response 3.3

Such selectivity is achieved as a result of the fluorescence properties of the analytes. In any given solvent system, only certain analyte molecules will fluoresce. It is therefore necessary to induce fluorescence by using light of a particular wavelength, with even fewer analytes then fluorescing at each of the many wavelengths available. When molecules fluoresce, they emit light at certain wavelengths only. By using a combination of excitation and emission wavelengths, and comparing these to known standards, it is possible to detect and identify specific molecules under the chromatographic conditions being employed.

Response 3.4

Many molecules absorb light at 220 nm – this is in the region where a process known as 'end absorption' occurs. This behaviour is not specific to any one (or class of) molecule(s), in contrast to fluorescence detection, which, under the correct conditions, can be specific to one particular molecule.

Response 3.5

This is probably a combination of at least two processes, with the dominant ones being (i) ion-pairing, and (ii) ion-exchange. In the former, the LSD is positively charged at pH 3.5, with the alkyl sulfonate acting as the counter-ion. The latter carries a lipophilic chain which will exhibit good chromatographic properties in the ODS stationary phase, so conferring good mass transfer behaviour. This, in turn, improves the chromatographic behaviour of the LSD alkyl sulfonate species. In addition (process (ii)), it is possible that the alkyl sulfonate partitions into the stationary phase, thus providing ion-exchange sites, where the sodium ions and positively charged LSD molecules (since the pH of the mobile phase is 3.5) compete for the active sites.

Response 3.6

These include resolution of those compounds which separate with similar retention times, for example, (i) iso-lysergic acid and dihydroergotamine, and (ii) LSD and N-methyl-N-propyl lysergamide. However, it is possible to overcome these problems by either changing the stationary phase, derivatization and/or the use of temperature programming.

Response 3.7

This number of scans were recorded in order to increase the signal-to-noise ratio in the spectrum. Since only very small amounts of LSD are present on a blotter acid, this is necessary to achieve a good quality spectrum for definitive identification. The advantage of using this particular scan range is that data collected between 700 and 1500 cm^{-1} are unique to LSD, since this represents the 'fingerprint region' of the spectrum for this compound. Collection of further data between 1500 and 4000 cm^{-1} will provide additional spectroscopic information.

Chapter 4

Response 4.1

This is because the item may need to be re-examined at a later stage. It may also be necessary, within the adversarial legal framework, for the 'other side' to examine the specimen. Sufficient material should therefore be left to allow for equal opportunity for re-examination by another party.

Response 4.2

While the non-polar cannabinoids are soluble and stable in diethyl ether, the insolubility of the carboxylic acids means that ether solutions cannot be used in sample preparation for comparative purposes. As a result, it will not be known what proportion of the sample has dissolved prior to analysis, or whether 'like is being compared with like'.

Response 4.3

It is not possible to determine whether all or some of the sample has broken down under such conditions, and hence whether different samples have degraded to the same extent. The consequence of this is that, again, sample comparison is not possible.

Response 4.4

We will use as an example, Δ^9-tetrahydrocannabinol, which has the following structure:

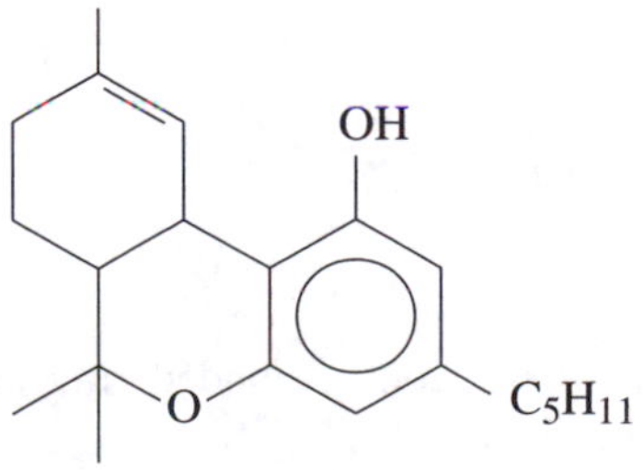

In this case, a number of different interactions with the silica gel can take place, including the following:

1. The free lone-pair electrons of the ether moiety hydrogen-bonding to the silanol group on the silica gel:

2. The free lone-pair electrons on the phenolic moiety hydrogen-bonding to the proton of the silanol group on the silica gel:

3. The proton of the phenolic moiety hydrogen-bonding to the oxygen of the silanol group on the silica gel:

4. The π-electron clouds of the double-bonded ring unit hydrogen-bonding to the silanol group on the silica gel:

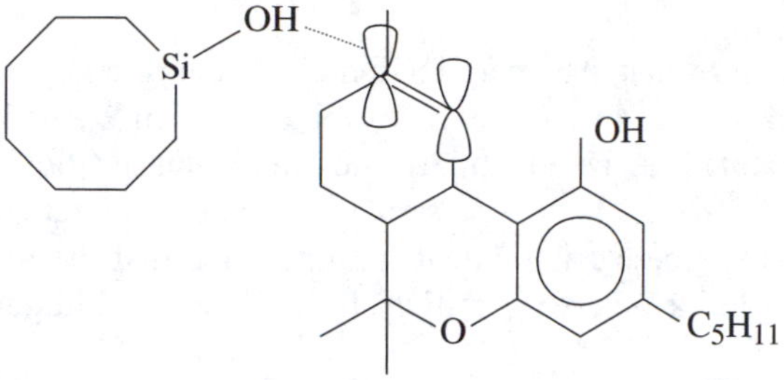

OH

Si

OH

O   C5H11

The stronger the interaction between the analyte and the cannabinoid, then the more strongly the material will be retained on the silica gel, thus resulting in a lower R_f value.

Response 4.5

This technique allows definitive identification because under a given set of conditions, the analytes will break down in a unique and predictable way. While there will be some variation in the mass spectra of an individual compound, even for the 'same' peaks within different spectra (as shown here for Δ^9-THC), the spectra obtained are still identifiably different to those of the very closely related Δ^8-THC compound.

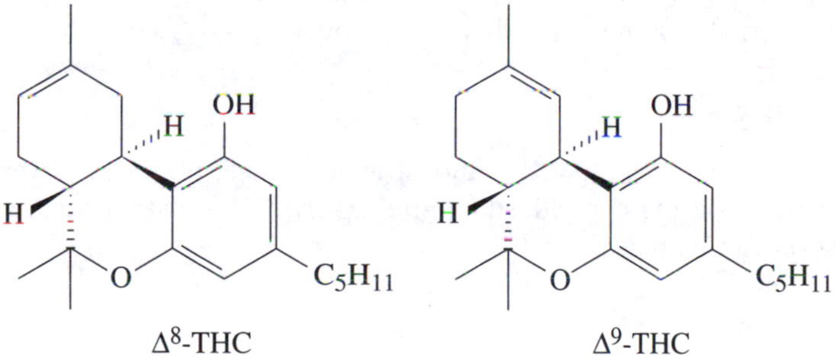

Δ^8-THC Δ^9-THC

It is this ability to discriminate between compounds that makes mass spectrometry such a powerful technique.

Chapter 5

Response 5.1

Acetic anhydride is an excellent acetylating reagent which reacts with free hydroxide groups in high yield, in this case to form diamorphine.

Response 5.2

This is to make sure that the material which is being analysed represents the *whole bulk* of the drug. In such a case, all of the different drug components present in a mixture can be identified and any quantification carried out will result in numerical data which represent the whole drug sample. Furthermore, if comparisons are to be carried out, it is imperative that the materials analysed represent the whole, or otherwise critical links between different samples could be missed.

Response 5.3

By using this chromatographic technique, we can determine (i) which members of the drug class are present and (ii) whether the samples can be discriminated on a qualitative basis. On the basis of such information, decisions can then be made as to whether investigations should proceed further and if so, which analytical techniques should be employed.

Response 5.4

There are a number of reasons for this. For example, in many of the solvent systems which are used, compounds are often not completely resolved, as exemplified by the data presented in Table 5.2. Furthermore, neither potassium iodoplatinate or Dragendorff's reagent are specific to opiates. It is therefore necessary to employ a more definitive technique.

Response 5.5

The blanks are needed in order to demonstrate that there is no carry-over between sample runs, while the derivatized samples provide the data against which the identification can be made.

Response 5.6

This is because there will have been chemical changes to the drug prior to its analysis which will result in changes in the mass spectra. This is illustrated in Figure SAQ 5.6 for 6-*O*-monoacetylmorphine and its BSA derivative. In addition, a change in the retention time will be observed in the gas chromatograms due to the change in molecular weight, boiling point and chromatographic behaviour of the derivatized drug component.

Response 5.7

Since the same batch of mobile phase is being used for both the test sample and the standard, their UV spectra should be identical, i.e. with respect to the

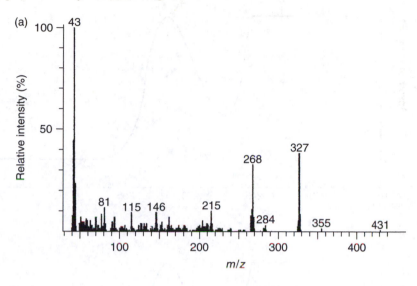

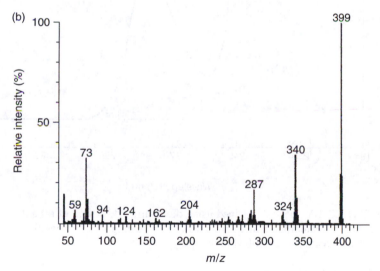

Figure SAQ 5.6 Electron-impact mass spectra of (a) 6-*O*-monoacetylmorphine, and (b) its trimethylsilyl ether derivative (after reaction with N,O-BSA), following gas chromatographic separation.

wavelengths where maximum and minimum absorptions occur (plus, if present, the positions of any shoulder features). If this is the case, as shown in Figure SAQ 5.7 for the HPLC analysis of a heroin (diamorphine) sample, then an identification of the analyte (drug sample) can be inferred.

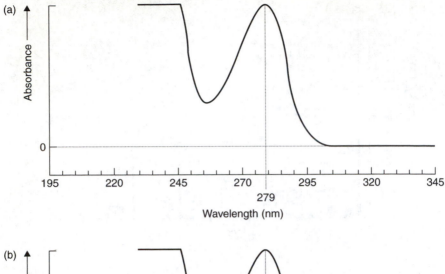

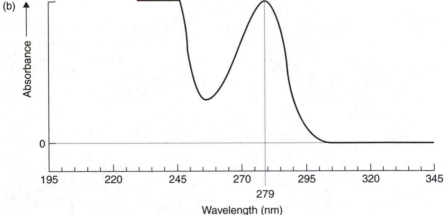

Figure SAQ 5.7 UV spectra, obtained using diode-array detection, of (a) a diamorphine standard, and (b) diamorphine in a heroin sample, following HPLC separation employing the system and conditions described in the text.

Chapter 6

Response 6.1

The base reacts with the sulfuric acid that is produced when the salt form of the drug dissociates in aqueous solution. The free base form of the drug which is subsequently generated is not freely soluble in water, and thus will precipitate.

Response 6.2

As with all tests involving any potential for 'contact' with the reagents being used, due care should be taken with respect to safety procedures. This includes wearing the correct protective clothing (e.g. safety glasses, gloves, etc.) and ensuring that there is no contact between the operator and the test materials. Any odours resulting from the latter should be tested from a safe distance away from the reaction mixture, i.e. by gently wafting the vapour towards the nose of the operator.

Response 6.3

This is a quality control procedure to ensure that there is no carry-over between the samples being tested.

Response 6.4

The principle reason is that, unlike HPLC, the GC analytical process does not involve the use of an injection loop in the instrumental system. This means that there will be an inherent variation between the volumes that are injected into the chromatograph. The use of internal standards provides a reference point since if the concentration is standardized, then the response of the detector to an analyte will be proportional to the amount injected.

Response 6.5

This is necessary because if different solutions were used, there would be an increase in experimental error, since the concentrations of the *internal* standard in the sample under investigation and in the standard, i.e. known, sample, would not be identical.

Chapter 7

Response 7.1

In this process, the hydrochloride salts of the drugs are returned to their free base forms (the acid reacts with the reagent being used to achieve basification) and the drugs thus become more lipophilic. The free bases will then redissolve in the organic solvent, being removed from, for example, some of the sugars, etc., which will have been extracted during the first part of the extraction procedure.

Response 7.2

This is necessary because any traces of water in the system would interfere with the quality of the chromatographic analysis. Removal of water from such solutions can be easily carried out by adding using a small amount of anhydrous sodium sulfate.

Response 7.3

The use of ninhydrin as a reagent provides a rapid and sensitive method for the detection of the primary amine groups in these alkaloids. Such amino groups react with this visualization reagent to form an iso-indole species, known as 'Ruhmann's purple'.

Response 7.4

This is required for a number of reasons, including (i) an 'illustration' of the material that is being analysed (e.g. for potential 'court-going' purposes), (ii) possible opportunities to identify the material botanically, and (iii) provision for carrying out identical analyses to those first performed.

Response 7.5

The ammonia in the solvent system will convert the mescaline present into its free base form, thus resulting in a more efficient extraction. Ether and methanol are immiscible, with the lipids being more soluble in the former. They are thus removed from the sample, so maintaining the drug component in the methanol (in which it is more soluble) as its free base form.

Response 7.6

The ammonia maintains the drug in its free base form through the process of ion-suppression. This is particularly important for the amino group in such a molecule, as this will become highly polar if salt formation takes place. The latter would result in considerable tailing and poor chromatographic behaviour because of the strong bonding that would result from interactions with the silanol groups on the silica gel plate. The ammonia stops the formation of the salt, reduces tailing and improves the overall chromatographic behaviour.

Chapter 8

Response 8.1

Fungal material contains a highly variable amount of water. This will affect the weight of the bulk sample, but *not* the amount of drug present. If the material is very wet, then the percentage amount of drug present, if measured in this condition, will be reported low. By drying the material, the variation in accuracy of the data due to the presence of water is removed and the results become comparable.

Response 8.2

The main problem here is that of ensuring that the blender is property cleaned after use. During such homogenization treatment, material can become trapped, for example, either under the blades or in the seals – both of these being areas which are difficult to clean effectively.

Response 8.3

In a reversed-phase system as in this example, when a number of analytes of different polarities are being investigated, some of these may be retained more strongly than others due to partitioning into the stationary phase. In order to elute such components from the column, the lipophilicity of the mobile phase must be increased, in this case by changing the proportion of acetonitrile in the solvent system. With increasing acetonitrile content, the polarity of the mobile phase will decrease and thus increasingly lipophilic compounds will elute.

Response 8.4

Although gradient systems have been employed for the HPLC analysis of a wide variety of drug samples, their use does have the major disadvantage that a long equilibration time is required between each determination. This is because the solvent contained in the pores on the surface of the silica gel stationary phase must be allowed to return to equilibrium with the bulk of the mobile phase at the starting composition.

Chapter 9

Response 9.1

This results from the difference in polarity of the substituent group at the C5 position of the barbiturate ring. Phenobarbitone contains a benzene ring, with a polar cloud of delocalized π-electrons, while cyclobarbitone contains a six-membered ring with an alkene group. In the latter compound, the resulting π-electron cloud will be less polar in character. Since the separation mechanism is principally sorption–desorption, the more polar phenobarbitone will be more strongly retained by the silanol groups on the silica gel, thus giving rise to a lower R_f value.

Response 9.2

This is needed because of the mode of separation, with reversed-phase HPLC being based on a partitioning process. As the lipophilicity of the side-chain at the C5 position increases, so the barbiturate compound will partition preferentially into the stationary phase, but not re-enter the mobile phase. Increasing the proportion of acetonitrile in the solvent system will result in better mass transfer characteristics, and hence improved chromatography, for these more lipophilic barbiturates.

Bibliography

Journals

The following journals frequently contain articles relating to the analysis of drugs:

- *Bulletin on Narcotics*
- *Forensic Science International*
- *Forensic Science Review*
- *Journal of Forensic Sciences*
- *Problems of Forensic Science* (*Z Zagadnień Nauk Sądowych*)
- *Science and Justice*

In addition, relevant articles may occasionally be found in the following periodicals:

- *Journal of Chromatography*
- *Journal of the Association of Official Analytical Chemists*
- *The Analyst*

Books

Legislation

The following are texts related to the Misuse of Drugs Act(s) and associated legislative documents. Although the two editions of 'Fortson' treat the material in different ways, both are useful, depending on the interest of the reader.

Fortson, R., *Misuse of Drug and Drug Trafficking Offences*, 3rd Edn, Sweet and Maxwell, London, 1996.

Fortson, *Misuse of Drug and Drug Trafficking Offences*, 4th Edn, Sweet and Maxwell, London, 2002.

King, L. A., *The Misuse of Drugs Act: A Guide for Forensic Scientists*, The Royal Society of Chemistry, Cambridge, UK, 2003.

Drug Analysis

Cole, M. D. and Caddy, B., *The Analysis of Drugs of Abuse – An Instruction Manual*, Ellis Horwood, Chichester, UK, 1995. Based on an experimental approach, this text is best used while undergoing practical training in the laboratory.

Gough, T. A., *The Analysis of Drugs of Abuse*, Wiley, Chichester, UK, 1991. A very useful reference text, although now a little dated in terms of currently available technology.

Moffat, A. C., Jackson, J. V., Moss, M. S., Widdop, B. and Greenfield, E. S., *Clark's Isolation and Identification of Drugs in Pharmaceuticals, Body Fluids and Post-Mortem Material*, The Pharmaceutical Press, London, 1986.

Siegel, J. A., Saukko, P. J. and Knupfer, G. C. (Eds), *Encyclopedia of Forensic Sciences*, 3-Volume Set, Academic Press, New York, 2000.

White, P. C. (Ed.), *Crime Scene to Court: The Essentials of Forensic Science*, The Royal Society of Chemistry, Cambridge, UK, 1998. Chapters 9 and 10 are of particular relevance, with both providing good general introductory material.

In addition, the United Nations Division of Narcotic Drugs (United Nations Drug Control Programme (UNDCP)) have published a number of *Manuals for Use by National Narcotics Laboratories* which give details of the recommended testing methods for various drugs and drug classes. These include the following:

- *Barbiturate Derivatives* (1989)

- *Benzodiazepine Derivatives* (1988)

- *Heroin* (1988)

- *Lysergic Acid Diethylamide (LSD)* (1989)

- *Peyote Cactus (Mescal Buttons)/Mescaline and Psilocybe Mushrooms/Psilocybin* (1989)

Chromatographic Techniques and Data Interpretation

Barker, J., *Mass Spectrometry*, 2nd Edn, ACOL Series, Wiley, Chichester, UK, 1999.

Blau, K. and Halkett, J. H., *Handbook of Derivatives for Chromatography*, Wiley, Chichester, UK, 1993.

Braithwaite, A. and Smith, F. J., *Chromatographic Methods*, Blackie Academic and Professional, London, 1996.

Chapman, J. R., *Organic Mass Spectrometry*, 2nd Edn, Wiley, Chichester, UK, 1995.

Fowlis, I. A., *Gas Chromatography*, 2nd Edn, ACOL Series, Wiley, Chichester, UK, 1995.

Hahn-Deinstrop, E., *Applied Thin Layer Chromatography: Best Practice and Avoidance of Mistakes*, Wiley-VCH, Weinheim, Germany, 2000.

Lindsay, S., *High Performance Liquid Chromatography*, 2nd Edn, ACOL Series, Wiley, Chichester, 1992.

McMaster, M. C. and McMaster, C., *GC/MS: A Practical User's Guide*, Wiley, New York, 1998.

Miller, J. C. and Miller, J. N., *Statistics and Chemometrics for Analytical Chemistry*, 4th Edn, Prentice Hall, Upper Saddle River, NJ, 2000.

Sherma, J., 'Thin Layer Chromatography', in *Encyclopedia of Analytical Chemistry*, Vol. 13, Meyers, R. A. (Ed.), Wiley, Chichester, UK, 2000, pp. 11 485– 11 498.

Web-Based Resources[†]

There are a large number of websites relating to drugs (of abuse). However, caution should be exercised when using material gained from such resources, e.g. the scientific rigour of the information should always be questioned. A recent review of websites which contain material on the clandestine use of drugs is available in the following:

Bogenschutz, M. P., 'Drug information libraries on the Internet', *Journal of Psychoactive Drugs*, **32**, 249–258 (2000).

Many websites are concerned with the clandestine use of drugs, as well as containing reliable scientific information. Of these, perhaps the most useful are the following:

- www.lycaeum.org
- www.erowid.com

The United Nations Drug Control Programme (UNDCP) also has a particularly useful website (www.undcp.org) from where a large amount of interesting material and articles can be accessed and also downloaded. An on-line version of the *Bulletin on Narcotics* is also available from this site, and in addition, links provide access to the relevant international legislative documents and facilities for downloading these.

[†] As of May, 2002. The material displayed is not endorsed by the author or the publisher.

Glossary of Terms

This section contains a glossary of terms, all of which are used in the text. It is not intended to be exhaustive, but to explain briefly those terms which often cause difficulties or may be confusing to the inexperienced reader.

Adnation Attachment of one organ to another by its whole length. A term used in botany to describe the gills of agarics (mushrooms and toadstools) (*see also* Adnexed *and* Sinuate).

Adnexed Joined by a narrow margin. A term used in botany to describe the gills of agarics (mushrooms and toadstools) (*see also* Adnation *and* Sinuate).

Adulterants Bulking agents, which are pharmacologically active, added to a drug to 'make it go further' and also mask the desired effect.

Analogues Compounds which are structurally related and derived from each other.

Analyte The material to be quantified, characterized or in any other way investigated.

Anorectic An appetite suppressant.

Ballistics The physical marks and characteristics of a tablet.

Base peak The most intense ion in a mass spectrum. The intensities of other ions in the spectrum are reported as a percentage of the intensity of this peak.

Blanks Materials which are analysed between (drug) samples, used to demonstrate that the equipment, instruments, etc. are 'clean' before analysis of the next sample.

Buffer An electrolyte added to an HPLC mobile phase.

Campanulate Bell-shaped.

Capillary column This term refers to a chromatographic column of 'small' diameter, used in both gas and high performance liquid chromatography. In HPLC, the term is usually applied to columns with internal diameters between 0.1 and 2.0 mm. The term 'microbore' column is often used synonymously to describe these columns, but is more correctly applied to those with internal diameters of 1 or 2 mm.

Check standard   A standard analysed in a sequence which allows the performance of an instrument to be monitored.

Chiral separation   The separation of chiral (e.g. optically active) molecules, for instance, the (*R*)- and (*S*)-isomers of amphetamines (*see also* Racemic mixture).

Chromaphore   That part of a molecule which absorbs light.

Chromatographic selectivity   The degree to which compounds are separated on a particular chromatographic system.

Column priming   The retention of analytes on a chromatographic column which can be observed in subsequent analyses.

Confirmatory testing   The process of testing to prove the *identify* of a drug.

Cutting agents   Materials added to a drug sample to increase its bulk (*see also* Adulterants).

Derivatization   The process of chemical modification of an analyte.

Diluents   (*see* Adulterants)

Diode-array (UV) detector   A detector which monitors all wavelengths simultaneously and therefore allows a complete UV spectrum to be obtained instantaneously. The alternative, a *dispersive* UV detector, monitors one wavelength at a time and thus requires a considerable amount of time to record a complete spectrum.

Dioecious   A plant species which has both male and female characteristics.

Entactogen   A compound which encourages physical contact.

Exhibit (production) label   An item which is used as evidence in court, or from which materials have been removed, possibly analysed and then used as evidence.

False positive   A positive response obtained in a test which results from a component other than the target analyte.

Flow programming   Varying the HPLC flow rate during the course of a separation.

Free base   A basic drug which has not reacted with an acid.

Gas chromatography–mass spectrometry   The combination of gas chromatography with mass spectrometry (*see also* Tandem technique).

Gradient elution   The changing of an HPLC mobile phase composition during the course of an analysis.

Heroin   The mixture of compounds produced when morphine is extracted from opium, purified and subsequently acetylated to diamorphine.

Hybrid technique   The combination of two or more analytical techniques (*see also* Tandem technique).

Hydrochloride salt   The compound formed following the reaction of a basic drug with hydrochloric acid.

Hyphenated technique   The combination of two analytical techniques (*see also* Hybrid technique *and* Tandem technique).

Injector   A common term used for the method of sample introduction into a chromatographic system.

Insufflacation   The process of drug ingestion through the nose.

Internal standard   A material added to a sample which is to be chromatographed, against which the retention time and detector response can be measured.

Ion-pairing reagent   A material which forms a complex with an ionic compound to allow its analysis using HPLC.

Lipophilicity   A measure of the degree of non-polarity; a lipophilic molecule is non-polar.

Mobile phase   That part of a chromatographic system which causes the analyte to move from the point of injection to the detector. In HPLC, this is a liquid.

Mycology   The study of fungi.

Negative control   A sample known to be free of a drug component, used to determine that the result obtained is due to the drug sample rather than the analytical process itself.

Normal-phase HPLC   An HPLC system in which the mobile phase is less polar that the stationary phase.

Occluded solvent   A solvent found in a drug sample following crystallization, which can be either chemically bound to the crystal itself or trapped in the interstitial spaces of the crystal lattice.

Packed column   An HPLC column containing particles of inert material of typically 5 μm diameter on which the stationary phase is coated.

Phenotypic plasticity   Variation in the observed form of an organism (in this context, a drug-producing plant or fungus) which is due to the combined effects of genetic make-up and environment.

Physical fit   The 'perfect' fit of two or more pieces of material such that it is possible to confirm that they originated from the same item.

Phytochemistry   Study of the chemical processes and composition of plant life.

Positive control   A known sample used to demonstrate that an (analytical) test is working.

Precision   The closeness of replicate measurements on the same sample.

Presumptive test   A screening test used to determine which class or classes of drugs might be present in a sample.

Proficiency test   A test used to determine whether (or not) it is possible to obtain the expected answer.

Profiling   The analysis and comparison of materials in a drug sample used to establish its origin, route of synthesis and/or relationship with other samples.

Qualitative analysis   The analysis of a sample to determine the *identity* of any compounds present.

Quantitative analysis   The analysis of a sample to determine the *amount* of an analyte present.

Racemic mixture   A mixture of chiral molecules (*see also* Chiral separation).

Repeatability   The closeness of a set of measurements carried out by a single analyst on a single instrument within a narrow time interval using the same reagents.

Reproducibility   The closeness of a set of measurements carried out by a number of analysts on a number of instruments over an extended period.

Retention index   The ratio of the retention times of the test sample and an internal standard.

Reversed-phase HPLC   An HPLC system in which the mobile phase is more polar than the stationary phase.

Robustness   The ability of a method to withstand changes in reagents, equipment, operator, etc.

Salt form   The form produced when a basic drug has reacted with an acid *or* an acidic drug has reacted with a base.

Selected-ion monitoring   A technique in which a mass spectrometer is used to monitor only a small number of ions characteristic of the analyte of interest.

Selective detector   A detector which responds only to compounds containing a certain structural feature.

Selectivity   The ability to determine the analyte of interest with accuracy and precision in the presence of other materials.

Separation factor   (*see* Chromatographic selectivity)

Sinuate   Wavy-edged, with distinct inward and outward bends along the edges. A term used in botany to describe the gills of agarics (mushrooms and toadstools) (*see also* Adnation *and* Adnexed).

Stationary phase   That part of the chromatographic system with which the analytes interact and over which the mobile phase flows.

Striae   Linear marks or streaks on a surface.

Sulfate salt   The compound formed following the reaction of a basic drug with sulfuric acid.

Tailing   The process of prolonged retention in a chromatographic system which results in asymmetric peak shapes, e.g. in HPLC and GC analyses.

Tandem technique   A term used for the combination of two or more analytical techniques (*see also* Hybrid technique).

Thermal decarboxylation   The process of chemical loss of carbon dioxide from a molecule due to the influence of heat.

Thermal lability   The process of chemical breakdown of a molecule due to the influence of heat.

Total-ion-current trace   A plot of the total number of ions reaching a mass spectrometry detector as a function of the analysis time.

Trichome   Hair-like structures found on the surfaces of certain plants. In the case of *Cannabis* there are a number of different types, which are particularly important in its identification.

SI Units and Physical Constants

SI Units

The SI system of units is generally used throughout this book. It should be noted, however, that according to present practice, there are some exceptions to this, for example, wavenumber (cm^{-1}) and ionization energy (eV).

Base SI units and physical quantities

Quantity	Symbol	SI Unit	Symbol
length	l	metre	m
mass	m	kilogram	kg
time	t	second	s
electric current	I	ampere	A
thermodynamic temperature	T	kelvin	K
amount of substance	n	mole	mol
luminous intensity	I_v	candela	cd

Prefixes used for SI units

Factor	Prefix	Symbol
10^{21}	zetta	Z
10^{18}	exa	E
10^{15}	peta	P
10^{12}	tera	T
10^{9}	giga	G
10^{6}	mega	M
10^{3}	kilo	k

(continued overleaf)

Prefixes used for SI units (*continued*)

Factor	Prefix	Symbol
10^2	hecto	h
10	deca	da
10^{-1}	deci	d
10^{-2}	centi	c
10^{-3}	milli	m
10^{-6}	micro	μ
10^{-9}	nano	n
10^{-12}	pico	p
10^{-15}	femto	f
10^{-18}	atto	a
10^{-21}	zepto	z

Derived SI units with special names and symbols

Physical quantity	SI unit		Expression in terms of base or derived SI units
	Name	Symbol	
frequency	hertz	Hz	$1\ \text{Hz} = 1\ \text{s}^{-1}$
force	newton	N	$1\ \text{N} = 1\ \text{kg m s}^{-2}$
pressure; stress	pascal	Pa	$1\ \text{Pa} = 1\ \text{N m}^{-2}$
energy; work; quantity of heat	joule	J	$1\ \text{J} = 1\ \text{Nm}$
power	watt	W	$1\ \text{W} = 1\ \text{J s}^{-1}$
electric charge; quantity of electricity	coulomb	C	$1\ \text{C} = 1\ \text{A s}$
electric potential; potential difference; electromotive force; tension	volt	V	$1\ \text{V} = 1\ \text{J C}^{-1}$
electric capacitance	farad	F	$1\ \text{F} = 1\ \text{C V}^{-1}$
electric resistance	ohm	Ω	$1\ \Omega = 1\ \text{V}^{-1}$
electric conductance	siemens	S	$1\ \text{S} = 1\ \Omega^{-1}$
magnetic flux; flux of magnetic induction	weber	Wb	$1\ \text{Wb} = 1\ \text{V s}$
magnetic flux density; magnetic induction	tesla	T	$1\ \text{T} = 1\ \text{Wb m}^{-2}$
inductance	henry	H	$1\ \text{H} = 1\ \text{Wb A}^{-1}$
Celsius temperature	degree Celsius	°C	$1°\text{C} = 1\ \text{K}$
luminous flux	lumen	lm	$1\ \text{lm} = 1\ \text{cd sr}$
illuminance	lux	lx	$1\ \text{lx} = 1\ \text{lm m}^{-2}$

(*continued overleaf*)

Derived SI units with special names and symbols (*continued*)

Physical quantity	SI unit		Expression in terms of base or derived SI units
	Name	Symbol	
activity (of a radionuclide)	becquerel	Bq	$1 \text{ Bq} = 1 \text{ s}^{-1}$
absorbed dose; specific energy	gray	Gy	$1 \text{ Gy} = 1 \text{ J kg}^{-1}$
dose equivalent	sievert	Sv	$1 \text{ Sv} = 1 \text{ J kg}^{-1}$
plane angle	radian	rad	1^a
solid angle	steradian	sr	1^a

[a] rad and sr may be included or omitted in expressions for the derived units.

Physical Constants

Recommended values of selected physical constants[a]

Constant	Symbol	Value
acceleration of free fall (acceleration due to gravity)	g_n	$9.806\,65 \text{ m s}^{-2}$[b]
atomic mass constant (unified atomic mass unit)	m_u	$1.660\,540\,2(10) \times 10^{-27} \text{ kg}$
Avogadro constant	L, N_A	$6.022\,136\,7(36) \times 10^{23} \text{ mol}^{-1}$
Boltzmann constant	k_B	$1.380\,658(12) \times 10^{-23} \text{ J K}^{-1}$
electron specific charge (charge-to-mass ratio)	$-e/m_e$	$-1.758\,819 \times 10^{11} \text{ C kg}^{-1}$
electron charge (elementary charge)	e	$1.602\,177\,33(49) \times 10^{-19} \text{ C}$
Faraday constant	F	$9.648\,530\,9(29) \times 10^4 \text{ C mol}^{-1}$
ice-point temperature	T_{ice}	273.15 K[b]
molar gas constant	R	$8.314\,510(70) \text{ J K}^{-1} \text{ mol}^{-1}$
molar volume of ideal gas (at 273.15 K and 101 325 Pa)	V_m	$22.414\,10(19) \times 10^{-3} \text{ m}^3 \text{ mol}^{-1}$
Planck constant	h	$6.626\,075\,5(40) \times 10^{-34} \text{ J s}$
standard atmosphere	atm	$101\,325 \text{ Pa}$[b]
speed of light in vacuum	c	$2.997\,924\,58 \times 10^8 \text{ m s}^{-1}$[b]

[a] Data are presented in their full precision, although often no more than the first four or five significant digits are used; figures in parentheses represent the standard deviation uncertainty in the least significant digits.

[b] Exactly defined values.

The Periodic Table

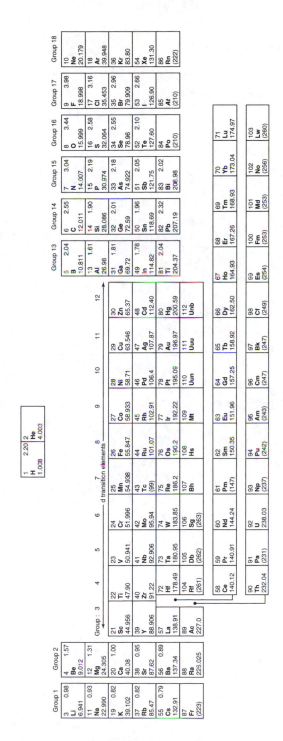

Index

Note: page numbers presented in italic script indicate an entry in the *Glossary of Terms*, while those presented in bold script refer to a detailed analytical method.

Acetic acid solvent, 70, 78, 142
Acetomenophen (paracetamol), 80, 81
Acetone solvent, 43, 144
Acetonitrile solvent, 45, 136, 149–150
Acetylcodeine, 77, 85, 91
Adnation (*see also* Adnexed and Sinuate), 129, *179*
Adnexed (*see also* Adnation and Sinuate), 129, *179*
Adulterants (*see also* Diluents), 14–15, 23, 80–82, 139, 140, *179*
Aminopropiophenone (cathinone), 114–115, 116–117, 118–119
Ammonia, 82, 103, 122, 123
Amphetamines, 13–36
　administration, 13–14
　adulterants, 14–15, 23
　classification, 15
　colour tests, 18, 121, 133, 158
　derivatization, 21–24
　dose, 14
　effects, 13
　extraction, **34**
　legislation, 3, 4, 6
　profiling, 31–35
　qualitative identification, 15–25
　quantification, 25–31
　ring-substituted, 20, 158

synthesis, 31–33
UK use, 2
Anaesthetics, 101, 103, 140
Analogues, *179*
Analytes, 83, 85, 150, *179*
Analytical process, 5–10, 153–155
Anorectic, *179*
Australian legislation, 5

Baeocystin, 128
Ballistic features, 16, 18, *179*
Barbitone, 139–140, 144, 145, 150
Barbiturates, 139–151
　colour tests, 80, 142–143, 157
　derivatization, 146
　extraction, 142
　legislation, 141
　qualitative identification, 142–145, 146–148
　quantification, 149–150
Base peak, 90–91, *179*
Benzodiazepines, 139–151
　colour tests, 80, 143, 159
　dose, 140
　extraction, 142
　legislation, 4, 6, 141
　qualitative identification, 143, 144, 146, 148–149

Benzoyl ecgonine, 98–100, 102, 103
Benzyl methyl ketone (BMK), 14, 32–33
N,*O*-Bis(trimethylsilyl)acetamide
  (*N*,*O*-BSA), 62–63, 67, 83–85,
  86–87, 104, 110
Blanks, 26, *179*
Blotter acid, 38–46
4-Bromo-2,5-dimethoxyphenethylamine
  (2C-B), 159
Buffer, 70, 123, *179*
Bulk analysis, 7–8, 53–54, 57–58, 61,
  78–79, 80, 104
Butobarbitone, 144, 145, 150

Cactus, 113, 119, 120, 123, 124
Caffeine, 15, 23–24, 27, 80, 81, 85
Campanulate, *179*
Cannabidiol (CBD), 51, 54–55, 61, 64,
  65, 70
Cannabinoids, 49–51, 52, 53, 58, 59–65,
  69–70
Cannabinol (CBN), 54–55, 61, 64, 66,
  70–71
Cannabis
  administration, 52
  colour tests, 58–59, 157–158
  comparison of samples, 65–71
  derivatization, 62–63, 67, 70
  dose, 54
  extraction, 59–61, 70
  herbal, 3, 50–51, 52, 54–57, 63, 67
  legislation, 3, 4, 6, 50–51
  qualitative identification, 54–71
  sources, 51–52
  UK use, 2
  wrapping materials, 51, 66–68
Cannabis oil (hash oil), 51, 52, 54, 57
Cannabis resin, 50–52, 54, 57, 68–69
Cannabis sativa (cannabis), 49–72
Capillary column, *179–180*
Carbon disulfide, 21–24
Catha edulis (*see also* Khat), 113–119,
  124–125
Cathine ((+)-norpseudoephedrine,(*S*,*S*)-
  norpseudoephedrine), 114–115, 116,
  118–119
Cathinone ((*S*)-aminopropiophenone),
  114–119

Check standard, *180*
Chiral separation (*see also* Racemic
  mixtures), 119, *180*
Chloroform solvent, 41–43, 79, 81–82,
  100, 102, 122, 124,
  134, 144
2-(2-Chlorophenyl)-2-(methylamino)-
  cyclohexanone (ketamine), 6,
  160
Chromaphore, *180*
Chromatographic selectivity, *180*
Cinnamoyl cocaines, 110
Classification of drugs, 3–4
Cobalt isothiocyanate, 80, 100–101, 157,
  158
Cocaine, 97–111
  administration, 99
  colour tests, 100–101, 157–159
  comparison of samples, 109–110
  derivatization, 104, 110
  effects, 99
  extraction, 99–100, 102, 104, 108–109
  legislation, 4, 6, 98
  qualitative identification, 100–107
  quantification, 107–109
  synthesis, 99–100
  UK use, 2
Coca leaf, 97–98, 100, 102
Coca paste, 98–100
Codeine
  legislation, 6, 74
  qualitative identification, 82–83, 85
  quantification, 91
  sources, 75–77
Colour tests, 18, 42–43, 58–59, 80,
  100–101, 121, 131–133, 142–143,
  157–159
Column priming, 90, *180*
Comparison of samples (*see also*
  Profiling), 65–71, 92–94, 109–110,
  119, 124
Cone-and-square method, **17–18**, 79, 121,
  154
Confirmatory testing, 20, 43–46,
  146–149, *180*
Contamination, 8, 79
Controlled Substances Act (USA), 5, 6

Corinth IV salt test, **58**
Court presentation, 9–10
Crack cocaine, 98, 99
Crimes (Traffic in Narcotic Drugs and
    Psychotropic Substances) Act, 1990
    (Australia), 5
Criminal Law Act, 1977, 50
Customs Act, 1901 (Australia), 5
Customs and Excise Management Act,
    1970, 3
Cutting agents (*see also* Adulterants), 8,
    151, *180*
Cyclobarbitone, 145
Cyclohexane solvent, 82, 103, 144

Dangerous Drugs Act, 1951, 3
Dangerous Drugs Act, 1964, 3
Dangerous Drugs Act, 1965, 3
Derivatization, 21–24, 62–63, 67, 83, 87,
    104, 110, 119, 133–135, 146, *180*
Designer drugs, 3
Diamorphine (*see also* Heroin), 73–95
    colour tests, 80, 101, 157
    legislation, 74
    preparation, 74–75, 77
    qualitative identification, 82, 85, 86
    quantification, 87–89, 90–93
Dille–Koppanyi test, 80, 142–143, 157
Diluents (*see also* Adulterants), 14–15, 81
2,5-Dimethoxy-4-(*n*)-
    propylthiophenethylamine (2C-T-7),
    160
3,6-Dimethyl-2,5-diphenylpyrazine, 116
Diode-array (UV) detector, 27–29, 90,
    118, 150, *180*
Dioecious, 50, *180*
DNA profiling, 71, 124, 136, 154–155
Dragendorff's reagent, 82, 103
Duquenois–Levine test, **58**, 157–158

Ecgonine, 98–100, 103, 104, 110
Ecgonine methyl ester, 98–100, 103
Ecstasy, 2
Ehrlich's reagent, 42–43, 132, 133, 158
Eicosane, 105–106
Electron-capture detection, 110
Electron-impact mass spectra, 21, 24, 65,
    86, 106

Entactogens, 13, *180*
Ephedrine, 3
Erythroxylum spp., 97, 100
Ester derivatives, 3, 4, 74
Ethanol solvent, 59–61, 102, 104
Ether derivatives, 3, 4, 62–66
Ethyl acetate solvent, 142
Euphoria (4-methylaminorex (4-methyl-5-
    phenyl-2-aminooxazoline)) (U4Euh),
    159
European Network of Forensic Science
    Institutes (ENFSI), 8
Evidence in court, 10
Exhibit (production) label, 115, *180*

False positive, *180*
Fast Blue BB, 60
Fischer–Morris test, 80
Flame-ionization detection, 85, 105, 107,
    110
Flow programming, 136, *180*
Flunitriazepam, 141
Fluorescence, 42–44, 122, 133, 146
Fourier-transform infrared (FTIR)
    spectroscopy, 68
Free base, *180*
Fungi, 37, 127–137

Gas chromatography (GC)
    cocaine, 107, 110
    heroin, 87–89
Gas chromatography–mass spectrometry
    (GC–MS) (*see also* Tandem
    technique), *180*
    amphetamines, 20–25, 34–35
    barbiturates, 146–148
    benzodiazepines, 148–149
    cannabis, 61–65, 70
    cocaine, 103–109
    heroin, 79, 83–87, 93–94
    Khat, 118–119
    LSD, 45
    mescaline, 124
    psilocin/psilocybin, 133–135
    retention times, 21–22, 24, 25, 35,
        63–64, 70, 85–86, 105, 118–119
Gradient elution, 136, 150, *180*

Hallucinogens, 37, 119, 127, 159, 160
Hash oil, 51, 52, 54, 57
Heptafluorobutyric anhydride (HFBA), 23,
    25, 110
Herbal cannabis, 3, 50–52, 54–57,
    63, 67
Heroin (*see also* Diamorphine), 73–95
    administration, 73–74
    adulterants, 80–82, 139, 140
    appearance, 77–78
    colour tests, 80
    comparison of samples, 92–94
    definition, 5, *180*
    derivatization, 83–85, 87
    dose, 73
    effects, 74
    legislation, 4–6, 73, 74
    paraphernalia, 78
    preparation, 77
    qualitative identification, 80–87
    quantification, 87–94
    solvent recovery, 79
    UK use, 2
Hexylamine, 123
High performance liquid chromatography
    (HPLC)
    amphetamines, **26–31**
    barbiturates, 149–150
    benzodiazepines, 150–151
    calibration curves, 26–31, 90–94
    cannabis comparison, 69–71
    heroin, 89–93
    Khat, 118
    LSD, 44–45
    mescaline, 122–124
    psilocin/psilocybin, 135–136
    solvents, 26, 44–45, 70, 89–90, 123,
        136
Homogenization, 17–18, 121, 131, 154
Hybrid technique, *180*
Hydrochloride salts, 77, 92, 98, 99, 117,
    180
Hyphenated technique, *181*

Impurities, 14–15, 31–35, 74, 154
Indole alkaloids (*see also* Lysergic acid
    diethylamide), 43, 158

Injector, *181*
Insufflacation, 13–14, 74, 99, *181*
Internal standard, *181*
Ionization energy, 183
Ion-pairing reagent, 44–45, 122, 123, *181*
Ion suppression, 20, 44, 69–70, 117–118
Ion-trap detectors, 106
Isocratic systems, 150

Kerosene, 99–100
Ketamine (2-(2-chlorophenyl)-2-
    (methylamino)-cyclohexanone), 6,
    160
Khat (*see also Catha edulis*), 113–119,
    124–125
    administration, 114
    comparison of samples, 119
    derivatization, 119
    effects, 114
    extraction, 117
    qualitative identification, 115–119
    wrapping materials, 114–116

Laboratory quality assurance, 9–10, 146
Least-squares method, 28, 88
Leuckart synthesis, 32
Licensing, 4
Lidocaine, 103
Lignocaine, 80, 82
Lipophilicity, 123, 150, *181*
Liquid–liquid extraction (LLE), **34**
Lophophora williamsii, 119–124, 125
LSD (*see* Lysergic acid diethylamide)
Lysergic acid, 6, 38
Lysergic acid diethylamide (LSD), 37–47
    dose, 38–41
    extraction, 41–42
    legislation, 6, 38
    qualitative identification, 38–46

Mandelin test, 80, 117, 158
Marijuana, 51
Marinol (synthetic tetrahydrocannabinol),
    6
Marquis test, 18, 80, 117, 121, 132–133,
    158
Mass spectrometry (MS) (*see* Gas
    chromatography–mass spectrometry)

Medicinal use, 4, 6, 140
Mercuric chloride–diphenylcarbazone
    reagent, 143–144
Mescaline
    (3,4,5-trimethoxyphenethylamine),
        119–120
    colour tests, 121
    comparison of samples, 124
    dose, 120
    extraction, 123–124
    legislation, 4, 119
    qualitative identification, 121–122
    quantification, 122–124
Methanol solvent
    extraction, 41, 79, 104, 108–109, 123,
        132, 133, 136, 142
    high performance liquid
        chromatography, 26–27, 44, 70,
        89, 90
    thin layer chromatography, 19, 43, 82,
        102–103, 122
Methaqualone, 80–82
4-Methylaminorex (euphoria) (U4Euh),
    159
Methylamphetamines, 13, 14, 20, 23, 27,
    28
    colour tests, 18, 133, 158
N-Methylation, 146, 148
Methyl benzoate odour test, 102
3,4-Methylenedioxyamphetamine (MDA),
    13, 14, 18, 20, 25, 27, 28
3,4-Methylenedioxyethylamphetamine
    (MDEA), 13, 14, 18, 20, 25
3,4-Methylenedioxymethylamphetamine
    (MDMA), 6, 13, 14, 18, 20, 23, 25,
    27, 28
Methylphenobarbitone, 141
N-Methyl-*N*-(trimethylsilyl)-2,2,2-
    trifluoroacetamide (MSTFA),
    134–135
Microscope FTIR spectroscopy, LSD,
    45–46
Misuse of Drugs Act, 1971, 2–5, 50–51,
    74, 98, 114, 119, 128
Misuse of Drugs Act, 1971 (Modification)
    Order 1986, 114

Misuse of Drugs Act (Regulations), 1985,
    2, 3, 98, 141
Misuse of Drugs Act (Regulations), 2001,
    2–4, 141
Mobile phase, 89, 118, 123, 150, *181*
6-*O*-Monoacetylmorphine, 77, 82, 83, 85,
    91
Morphine
    derivatization, 84
    diamorphine preparation, 73–77
    legislation, 4, 6, 74
    qualitative identification, 82–85
    quantification, 91
Mushrooms, 127–136
Mycology, 130, *181*

Negative control, 101, *181*
Ninhydrin, 117–118, 122
Nitrostyrene synthesis, 33
Norcocaine, 110
Norephedrine, 116, 118
Normal-phase HPLC, *181*
Norpseudoephedrine (cathine), 114–116,
    118–119
Noscapine, 75, 85, 91

Occluded solvent, 78, *181*
Offences, 4–5
Opiates
    colour tests, 18, 80, 158
    qualitative identification, 80–87
    quantification, 87–92
Opium, 3, 4, 6, 74–75

Packaging, 51, 66–68, 114–116, 130,
    131
Packed column, *181*
Papaverine, 75, 85, 91
Papaver somniferum (field poppy), 73,
    74
Paracetamol (acetomenophen), 80, 81
Pellotine, 120
Pentobarbitone, 144, 145, 150
Periodic table, 187
Personal protective equipment
    (PPE), 8
Peyote cactus, 113, 119, 120, 123, 124

Pharmaceutical drugs (*see also*
    Barbiturates and Benzodiazepines), 4,
    6, 70, 80, 139–151, 157, 159
Phencyclidine (PCP), 6, 159
Phenethylamines, 121
Phenobarbitone, 80, 139–140, 145,
    147–148
Phenolic compounds, 59, 60, 76
Phenotypic plasticity, 130, *181*
1-Phenyl-1,2-propanedione, 116
Photography, 16
Physical constants, 185
Physical fit, 54, 66, 68, 69, *181*
Phytochemistry, *181*
Piperocaine, 102
Positive control, 101, *181*
Powder samples, 15–16, 77, 79, 89, 102,
    115, 121
Precision, *181*
Prescription medicines (*see*
    Pharmaceutical drugs)
Presumptive tests, *181*
  amphetamines, 18
  cannabis, 58–59
  cocaine, 100–102
  colour, 18, 42–43, 58–59, 80,
      100–101, 121, 131–133, 142–143,
      157–159
  heroin, 80–81
  LSD, 42–43
  mescaline, 121
  pharmaceutical drugs, 142–143
  psilocin/psilocybin, 131–133
Procaine, 80, 82, 103
Proficiency test, *181*
Profiling (*see also* Comparison of
    samples), 31–35, 153, *181*
Psilocin, 127–137
  colour tests, 131–133, 158
  derivatization, 133–135
  qualitative identification, 129–136
  quantification, 135–136
  US legislation, 129
Psilocybe, 127, 129–130, 136
Psilocybin, 127–137
  colour tests, 131–133, 158
  dose, 128

extraction, 132–134, 136
  legislation, 6, 128, 129
  qualitative identification, 129–136
  quantification, 135–136

Qualitative analysis, *181*
  amphetamines, 15–25
  barbiturates, 142–145, 146–148
  benzodiazepines, 143, 144, 146,
      148–149
  cannabis, 54–71
  cocaine, 100–107
  codeine, 82–85
  diamorphine, 82, 85, 86
  heroin, 80–87
  Khat, 115–119
  LSD, 38–46
  mescaline, 121–122
  morphine, 82–85
  opiates, 80–87
  psilocin/psilocybin, 129–136
Quality assurance (QA), 9–10, 146
Quantitative analysis, *181*
  amphetamines, 25–31
  barbiturates, 149–150
  cocaine, 107–109
  codeine, 91
  diamorphine, 87–93
  heroin, 87–94
  mescaline, 122–124
  morphine, 91
  opiates, 87–92
  psilocin/psilocybin, 135–136
Quinalbarbitone (secobarbitone), 141

Racemic mixtures (*see also* Chiral
    separation), 119, *182*
Record keeping, 9–10
Regression equation, 28–31, 88–90, 92,
    108
Repeatability, *182*
Report writing, 9
Reproducibility, *182*
Research use, 4
Retention index, 85–86, 105–106, 134,
    182
Reversed-phase HPLC, 44, 69, 122, *182*
Robustness, *182*

Salt form (*see also* Hydrochloride salts and Sulfate salts), *182*
Sampling procedures, 153–154
  amphetamines, 15–18
  cannabis, 53–54
  heroin, 78–79
  Lophophora williamsii, 120–121
  LSD blotter acid, 39–41
  pharmaceutical drugs, 141–142
  psilocin/psilocybin, 131
Scott test, 101, 158–159
Secobarbitone (quinalbarbitone), 141
Selected-ion monitoring, *182*
Selective detector, *182*
Selectivity, *182*
Separation factor (*see* Chromatographic selectivity)
Sinuate (*see also* Adnation and Adnexed), 129, *182*
SI units, 183–185
Smell
  cocaine, 98, 102
  heroin, 77–78, 80
Solvents
  extraction, 41–42, 59–60, 61, 70, 79, 99, 102, 123, 132–134, 136, 142, 154
  high performance liquid chromatography, 26, 44–45, 70, 89–90, 123
  occluded, 78, *181*
  thin layer chromatography, 19, 43, 59–61, 81–82, 102–103, 122, 133
Stationary phase, 118, 123, *182*
Stereoisomers, 3, 4, 141
Stimulants, 13, 114, 159
Striae, 68, 115–116, *182*
Sulfate salts, 27, 99–100, *182*
Symbols, 183–185

Tabletted samples, 16–17
Tailing, *182*
Tandem technique, *182*
Temazepam, 101, 141, 157

Δ^9-Tetrahydrocannabinol (Δ^9-THC), 6, 50–52, 54, 61, 62–67, 70
Thebaine, 75
Thermal decarboxylation, 52, *182*
Thermal lability, *182*
Thin layer chromatography (TLC)
  amphetamines, **19–20**
  barbiturates, 143–145
  benzodiazepines, 144, 146
  cannabis, 59–61
  cocaine, 102–103
  heroin, 81–83, 93
  Khat, 117–119
  LSD, 43
  mescaline, 121–122
  psilocin/psilocybin, 133
  solvents, 19, 43, 59–61, 81–82, 102–103, 122, 144
Thiopentone, 141
Toluene solvent, 82, 103, 144
Total-ion-current trace, *182*
Trace analysis, 7–8, 17, 53, 58, 78–79, 104
Trichome, 55–57, *182*
3,4,5-Trimethoxyphenethylamine (*see* Mescaline)
Trimethylanilinium hydroxide, 146–147
Trimethylsilyl (TMS) ethers, 62–66, 134–135
Tropane alkaloids, 80, 99–100
Truxillines, 110

United Kingdom
  drug use, 1–2
  legislation, 2–5, 38, 50–51, 74, 98, 119, 128, 141
United Nations Convention against Illicit Traffic in Narcotic Drugs and Psychotropic Substances, 1988, 2
United Nations Convention on Psychotropic Substances, 1971, 2, 114, 141
United Nations Drug Control Programme (UNDCP), 15, 41, 78, 131, 141, 177
United Nations Single Convention on Narcotic Drugs, 1961, 2

United States legislation, 2, 5, 6, 38, 73,
    74, 98, 119, 129, 141
UV spectroscopy, 108–109

Van Urk's reagent (Ehrlich's reagent), 158
Veterinary use, 4

Wavenumber, 183
Web-based resources, 177
'Window panes', 38
Witness statements, 9

Zimmerman test, 143, 159